The Art of Teaching Physics with Ancient Chinese Science and Technology

中国古代科学技术与物理教学

Synthesis Lectures on Engineering, Science, and Technology

Each book in the series is written by a well known expert in the field. Most titles cover subjects such as professional development, education, and study skills, as well as basic introductory undergraduate material and other topics appropriate for a broader and less technical audience. In addition, the series includes several titles written on very specific topics not covered elsewhere in the Synthesis Digital Library.

The Art of Teaching Physics with Ancient Chinese Science and Technology
Matt Marone
2020

Nanotechnology Past and Present: Leading to Science, Engineering, and Technology
Deb Newberry
2020

Theory of Electromagnetic Beams
John Lekner
2020

The Big Picture: The Universe in Five S.T.E.P.S.
John Beaver
2020

Relativistic Classical Mechanics and Electrodynamics
Martin Land and Lawrence P. Horwitz
2019

Generating Functions in Engineering and the Applied Sciences
Rajan Chattamvelli and Ramalingam Shanmugam
2019

Transformative Teaching: A Collection of Stories of Engineering Faculty's Pedagogical Journeys
Nadia Kellam, Brooke Coley, and Audrey Boklage
2019

Ancient Hindu Science: Its Transmission and Impact on World Cultures
Alok Kumar
2019

Value Rational Engineering
Shuichi Fukuda
2018

Strategic Cost Fundamentals: for Designers, Engineers, Technologists, Estimators,
Project Managers, and Financial Analysts
Robert C. Creese
2018

Concise Introduction to Cement Chemistry and Manufacturing
Tadele Assefa Aragaw
2018

Data Mining and Market Intelligence: Implications for Decision Making
Mustapha Akinkunmi
2018

Empowering Professional Teaching in Engineering: Sustaining the Scholarship of Teaching
John Heywood
2018

The Human Side of Engineering
John Heywood
2017

Geometric Programming for Design Equation Development and Cost/Profit Optimization (with
illustrative case study problems and solutions), Third Edition
Robert C. Creese
2016

Engineering Principles in Everyday Life for Non-Engineers
Saeed Benjamin Niku
2016

A, B, See... in 3D: A Workbook to Improve 3-D Visualization Skills
Dan G. Dimitriu
2015

The Captains of Energy: Systems Dynamics from an Energy Perspective
Vincent C. Prantil and Timothy Decker
2015

Lying by Approximation: The Truth about Finite Element Analysis
Vincent C. Prantil, Christopher Papadopoulos, and Paul D. Gessler
2013

Simplified Models for Assessing Heat and Mass Transfer in Evaporative Towers
Alessandra De Angelis, Onorio Saro, Giulio Lorenzini, Stefano D'Elia, and Marco Medici
2013

The Engineering Design Challenge: A Creative Process
Charles W. Dolan
2013

The Making of Green Engineers: Sustainable Development and the Hybrid Imagination
Andrew Jamison
2013

Crafting Your Research Future: A Guide to Successful Master's and Ph.D. Degrees in Science &
Engineering
Charles X. Ling and Qiang Yang
2012

Fundamentals of Engineering Economics and Decision Analysis
David L. Whitman and Ronald E. Terry
2012

A Little Book on Teaching: A Beginner's Guide for Educators of Engineering and Applied Science
Steven F. Barrett
2012

Engineering Thermodynamics and 21st Century Energy Problems: A Textbook Companion for
Student Engagement
Donna Riley
2011

MATLAB for Engineering and the Life Sciences
Joseph V. Tranquillo
2011

The Art of Teaching Physics with Ancient Chinese Science and Technology
Matt Marone

ISBN: 978-3-031-00960-0 print
ISBN: 978-3-031-02088-9 ebook
ISBN: 978-3-031-00160-4 hardcover

DOI 10.1007/978-3-031-02088-9

A Publication in the Springer series
SYNTHESIS LECTURES ON ENGINEERING, SCIENCE, AND TECHNOLOGY
Lecture #10

Series ISSN 2690-0300 Print 2690-0327 Electronic

The Art of Teaching Physics with Ancient Chinese Science and Technology

中国古代科学技术与物理教学

Matt Marone
Mercer University

SYNTHESIS LECTURES ON ENGINEERING, SCIENCE, AND TECHNOLOGY #10

ABSTRACT

Blending physics with the study of ancient Chinese science, technology, and culture is a unique and highly effective way to present the fundamentals of physics to non-science majors. Based on the author's course at Mercer University (Georgia, U.S.), *The Art of Teaching Physics with Ancient Chinese Science and Technology* exposes a wide range of students to the scientific method and techniques of experimental analysis through the eyes and discoveries of ancient Chinese "polymaths" long before the European concept of the scientific method was even considered. No other book so deftly makes the connections from ancient China to Ben Franklin to Michael Faraday while teaching physics at the same time.

A distinctive characteristic of this book is the detailed hands-on laboratory experiments. This first includes making a simple magnetic compass and magnetometer. Students then use the compass/magnetometer to measure the strength of the magnetic field produced by a long straight wire. The second experiment covers two different methods of mining copper to introduce students to simple chemical principles such as displacement reactions, oxidation, reduction, and electronegativity.

Originally developed for non-science students in an Asian studies environment, this book provides a valuable resource for science teachers who wish to explore the historical connections largely ignored in traditional texts. When paired with *Teaching Physics through Ancient Chinese Science and Technology* (Marone, 2019), these two texts provide a unique means of studying selected topics traditionally found in a two-semester Physics course.

KEYWORDS

magnetic fields, compass, 罗盘 luópán, Michael Faraday, Benjamin Franklin, 沈括 Shěn Kuò, 梦溪笔谈 Mèng Xī Bǐtán, Brush Talks from Dream Brook, Spherical Mirrors, 墨子 Mòzǐ, Wet copper mining, malachite

Dedication

To polymaths the world over.

Contents

Preface

This is my second book on the subject of using ancient Chinese science as a means of teaching physics. In the first book, *Teaching Physics through Ancient Chinese Science and Technology*, the topics were mainly concerned with the study of mechanics and motion. This latest work takes up a few topics more commonly found in the second or third semester of general physics. We begin our study with magnetism. Magnetism has been known for thousands of years and has been studied by many different cultures. Our study of magnetism starts with the writings of the Song dynasty (宋朝 Sòng cháo) polymath 沈括 Shěn Kuò. In his work 梦溪笔谈 Mèng Xī Bǐ Tán, or *Brush Talks from Dream Brook*, 沈括 Shěn Kuò tells us how to make a compass. We also study the experiments of Franklin, Faraday, Henry, and Ørsted. The pioneering work of these scientists led to the unification of electricity with magnetism and in turn made possible the modern technologies we all now take for granted.

In our study of mechanics we generally deal with contact forces but magnetic and electric forces are exerted on objects that are not in physical contact with one another. This is an example of "action at a distance" and led to the concept of fields. including gravitational, electric, magnetic, etc.. In studying ancient concepts about magnetism, one cannot help but see how these interactions could be characterized as mystical or magical. The concept of 气 qì in an integral part of Chinese, or more broadly Asian, understanding of the cosmos. The connection between magnetism and the flow of 气 qì naturally brings up the concept of 风水 fēngshuǐ.

The optical properties of spherical mirrors were well known in ancient China. During the Waring States period, 战国时代 Zhànguó Shídài (476–221 BC), a Chinese work known as the 墨子 Mòzǐ, was published. The book is often studied in classes on ancient Chinese philosophy. In addition to philosophical discussions of ethics, governance, and economics, there are several sections that treat technology. Here we find an early discussion of images formed by spherical mirrors. In the Chapter 2, we introduce the concepts of ray tracing and the mirror equation. With these physical principles in hand we can then model and compare the image properties we calculate to those described in the 墨子 Mòzǐ.

What would our modern world be like if we had not discovered metals? Chapter 3 treats the basic properties of metals. Bronze is the alloy that changed the world. In this chapter we look into the concepts behind alloys and discuss some aspects of chemistry. Bronze consists chiefly of copper, and the mining of copper in ancient China is a topic of special interest. We introduce the technique of "wet copper" mining and study the chemistry behind the process.

One distinctive characteristic of this method of teaching physics is the emphasis on laboratory experiments. In this book, we have included experiments on magnetism and copper mining. In Chapter 4, following the suggestions of 沈括 Shěn Kuò, students build their own magnetic compass. We then turn that compass into a magnetometer and measure the magnetic field produced by a long straight wire. Our objective is to examine how the magnetic field depends on the current through the wire and the distance away from the wire. Students are introduced to several graphical techniques that enable them to determine the proportionalities expressed in the theoretical equation for the magnetic field.

In addition, we have also included two experiments related to copper mining which are described in Chapter 5. In one experiment we learn about smelting. Malachite is a common stone used for making jewelry. What few people realize is that it is a copper carbonate and can be smelted to produce copper. We present a laboratory experiment that is analogous to the method used in ancient times. "Wet copper" mining is a low-cost method of copper mining that introduces some simple chemistry. A solution of copper sulfate can be "mined" by allowing the liquid to contact iron or steel. This illustrates a simple displacement reaction and introduces the concept of electronegativity. The wet copper method was used in ancient times and small-scale operations can still be found in modern China.

The overarching theme in this method of teaching is to start with a discovery or invention from ancient China. We discuss how the ancients understood the phenomena in their world view and then analyze it based on the principles of physics. We have found that this method is particularly helpful for the non-science student who may be apprehensive about studying physics. Throughout this book we use authentic Chinese terminology. Chinese words are rendered in simplified characters with the 汉语拼音 Hànyǔ Pīnyīn system of romanization. Tone marks are given as an aid to pronunciation. We follow the system commonly used in the People's Republic of China.

Cover

The object on the cover is known as a 罗盘 luópán. At its center is what looks like an ordinary magnetic compass with the cardinal directions in Chinese characters. You might think that the first use of a magnetic compass was as a nautical navigation device. Compasses were used by the Chinese for nautical navigation around the year 1000. However, the ancient Chinese inventors of the compass used it for 风水 fēngshuǐ purposes as far back at the 4th century BC. 风水 fēngshuǐ is sometimes translated as "geomancy" but this term tends to obfuscate the actual meaning and leads to all sorts of misunderstanding of the basic principle as understood by the ancient Chinese. The term contains two characters, 风 fēng (wind) and 水 shuǐ (water). The overall meaning of the term has to do with the ebb and flow of the nearly untranslatable concept of 气 qì. Thus, in some way, the ancient Chinese saw the 罗盘 luópán as a sort of 气 qì meter. The concentric rings known as 層 céng are encoded with information that allows the 风水 fēngshuǐ practitioner to conduct measurements and draw conclusions.

Upon the 罗盘 luópán sits a hairy black rock which attracts the needle. This mineral is called 磁石 císhí, which is also known as magnetite or lodestone. 磁石 císhí is an iron oxide (Fe_3O_4) and is a naturally occurring magnet. The magnetic properties of this mineral were known by both the ancient Chinese and the ancient Greeks since about 500 BC. We use the term "naturally occurring magnet," to highlight the fact that, prior to about 1820, magnetite was the primary material used to produce magnetic fields. Soon after the experimental connection of magnetism to electricity by Hans Christian Ørsted, electromagnets provided a new and powerful source of magnetic fields. As we all know, the Earth is a giant magnet and the 罗盘 luópán responds to variations in the local magnetic field. In this book, the 罗盘 luópán serves as a symbol representing the connection between the ancient Chinese understanding of the natural world and the science of electromagnetism.

Acknowledgments

There are so many wonderful people who helped to make this book possible. Once again, Mercer University granted me sabbatical leave to investigate the history of Chinese Science at the Needham Research Institute in Cambridge. I would like to extend special thanks to Dr. Anita Gustafson, Dean of the College of Liberal Arts and Sciences, who helped to make this sabbatical possible. As always, my friends at the Needham Research Institute (NRI) were so kind and hospitable. The Director of the NRI, Dr. Jianjun Mei (梅建军), is such an encouragement to me and to the young scholars that he hosts at the NRI. As one Chinese scholar put it "The NRI is the Holy Land." Part of the "Holy Land" experience is the direct result of the tireless efforts of John Moffet who brings us scholastic treasures. I also wish to thank Ms. Susan Bennett, Administrative Manager of the NRI, who keeps everything moving in the right direction. Shanjia Zhang (张山佳), a visiting scholar at the NRI, was especially helpful in enabling me to obtain a Chinese magic mirror for study. I am also indebted to The Museum of Arts and Sciences in Macon, Georgia for allowing me to photograph specimens in their excellent mineral collection.

One of the exciting aspects of writing a cross-disciplinary book like this is the opportunity to meet new colleagues. I extend special thanks to Dr. Alexander Jost (曹大龍), Senior Scientist in the Department of History at Salzburg University, for his assistance with the history of wet copper mining and his photographic skill.

My friend and mentor Dr. Ronnie Littlejohn, Professor of Philosophy and Director of Asian Studies at Belmont University, has been a constant source of encouragement and ardent supporter of blending the study of Physics with Asian Studies. None of my adventures in Chinese science would have been possible without the support of my wonderful wife and family. They have sacrificed their time and effort to enable my work on this project.

Introduction

In *Teaching Physics through Ancient Chinese Science and Technology*, we dealt mainly with mechanical systems. We discussed the idea of what science is and how it was framed in Ancient China long before the European concept of the scientific method. In this volume we will explore material more commonly found in a second or third semester of an introductory physics class. China had no shortage of polymaths. Unfortunately, they are virtually unknown in the western world. In this volume we will explore the writings of the Song dynasty (宋朝 Sòng cháo) polymath 沈括 Shěn Kuò. Once again we have included an experimental section which contains detailed instructions for a laboratory component. For readers not familiar with this method of teaching physics we should reiterate the overarching method we use.

First, we explore physical phenomena in their authentic historical Chinese context. How did the ancient Chinese understand the physical world in terms of their own cosmology? This leads us into philosophical ideas very different from what one might normally encounter in a western science classroom. Then, we examine the same phenomena through the lens of modern science seeking out the basic underlying physics. Finally, we perform some type of experiment or recreation employing modern laboratory techniques. This teaching style is largely aimed at non-science majors who have misapprehensions and even fears about taking a science class. Of course, the material presented can be adapted to a wide range of students and classroom environments. Instructors are encouraged to build upon the ideas presented here and tailor them to their own unique teaching environments.

In 2015, Dr. Marone gave a talk at Belmont University covering the Ancient Chinese Science and Technology class he teaches. You can view this at: https://www.youtube.com/watch?v=mPQ8BvAXI6U.

CHAPTER 1

Magnetism (磁学 Cíxué)

1.1 THE MAGNETIC COMPASS 指南针 ZHǏNÁNZHĒN

Ask any educated Chinese person and they will tell you about the Four Great Chinese Inventions (四大发明 sìdàfāmíng). It might come as a shock to some in the West but the Chinese are credited with the invention of the magnetic compass (指南针 zhǐnánzhēn), papermaking (造纸 zàozhǐ), printing (木版 mùbǎn), and gunpowder (火药 huǒyào). These are by no means the only or even the most impressive of the long list of Chinese inventions. They are, however, inventions that have changed the world. Francis Bacon, considered by some to be the father of the Scientific Method, pondered the importance of three of these inventions. In the early 1600's he published the *Novum Organum* (Bacon, 1960). While musing over the power and results of discoveries he reminds us of the following.

> Again, it is well to observe the force and virtue and consequences of discoveries, and these are to be seen nowhere more conspicuously than in those three which were unknown to the ancients, and of which the origin, though recent, is obscure and inglorious; namely, printing, gunpowder, and the magnet. For these three have changed the whole face and state of things throughout the world; the first in literature, the second in warfare, the third in navigation; whence have followed innumerable changes, insomuch that no empire, no sect, no star seems to have exerted greater power and influence in human affairs than these mechanical discoveries.

The comment linking the magnet and navigation is where we begin our study of an ancient Chinese technology—the compass. We might automatically assume that the compass has always been a tool for navigation, but we shall soon see that in ancient China it had some surprising uses. When we think of a compass, it might be better to think of it as a device to measure the horizontal component of the Earth's magnetic field. Generally, we call a device that measures magnetic field strength a magnetometer. One of the earliest descriptions of a magnetic compass and magnetic declination comes to us from the brush of the famous Song dynasty (宋朝 Sòng cháo) polymath 沈括 Shěn Kuò. In his work 梦溪笔谈 Mèng Xī Bǐtán or *Brush Talks from Dream Brook*, 沈括 Shěn Kuò tells us how to make a compass and points out a rather strange observation. The translation here is my own and the original Chinese text can be found at the Chinese Text Project (https://ctext.org).

> Adepts can make a needle point to the south by rubbing it with a magnetic stone. However, the needle often slants to the southeast, not pointing due south. If the

needle is floating on water it is unsteady. The needle may be placed on a fingernail or the edge of a bowl, which will make it turn more easily, but since these supports are hard and smooth, it may easily fall off. The best way, is to suspend the needle by a single fiber of raw silk from a cocoon, which is attached to the center of the needle by a small piece of wax the size of a mustard seed. Then, hang it in a windless place and it will often point south. Needles may sometimes point North after being rubbed. In my house, I have needles that point North and others that point south. The south pointing property is like the habit of cypress trees that always lean to the west. Nobody knows the reason why it is so.

Remember that the overarching model in this method of teaching physics is to examine a technology used in ancient China and to pull out the underlying physical principles. We wish to explore how the Chinese understood or used the technology in their own cultural context and how we might understand in a physics classroom. Along the way we will also examine Chinese history and philosophy.

Returning to the statement about south pointing needles, let's see what we can learn. I have chosen to use the word "Adepts" in this translation for a very particular reason. The Chinese term is 方家 (fāngjiā) which modern translators might render as "expert" or "learned." Needham translates this as "Magicians" (Needham, Wang, and Robinson, 1962). As a noun, adept originally meant someone who has knowledge of alchemy, the occult, hermetic philosophy, and magic. Magnetism does appear magical in that it acts upon an object through a non-contact force. We often think of a force as being transmitted by some sort of push or pull that makes physical contact with the object upon which the force is acting. For centuries that mechanical contact force was the foundation of our understanding. When scientists began to study gravity, electrostatics, and magnetism we began to think about action at a distance taking place through some type of field.

The needle is rubbed with a magnetic stone. In Chinese this stone is called 慈石 císhí. In English, we call this rock lodestone or, more formally, the mineral magnetite. Figure 1.1 shows a piece of lodestone hanging from a thread.

Chemically, magnetite is composed of iron and oxygen. It consists mainly of Fe_3O_4 with inclusions of Fe_2O_3 and other impurities. Keep in mind that this is not the earliest reference to the magnetic properties of lodestones or the fact that they attract a needle. The interaction between a lodestone and needle was mentioned in the 吕氏春秋 Lǚshì Chūnqiū (Mr. Lü's *Spring and Autumn Annals*) dating back to about 239 BC and may also have appeared earlier. Readers familiar with Chinese can find this reference online at the following: https://ctext.org/lv-shi-chun-qiu/jing-tong. In this book we are not so much concerned with the question of priority. That is to say, questions regarding who first invented something or when did a technology transfer from one culture to another. We are more concerned with finding clear statements of fact that we can analyze in terms of the underlying physics. What we want to explore is that fact that the needle orients itself and that it does not align north–south. Why should it orient itself at all?

Figure 1.1: Lodestone 慈石 císhí hanging from a thread. Note "S" on right side indicating the south pointing end of the stone.

Notice the text tells us that the needle points south. If you are raised in the European tradition, you are inclined to think that a compass needle should point to the north not the south. I once remember reading about a person complaining that a compass they had purchased from China was broken because it pointed to the south not the north. Of course, if you have a needle and one end points north then the other end must point south. There is more to it than that. If you are in the northern hemisphere, as China is, go outside around midday and face south. Do you see the Sun? Now turn around and face north. Do you still see the Sun? If you sit with your back to the north and your face to the south you have a good view of the Sun throughout the day. The ancient Chinese gave importance to the southern direction for metaphysical reasons as we will see when we discuss the theory of Feng Shui (风水 fēngshuǐ).

The text tells us that the needle does not point due south but it slants or inclines a little to the east. At first this little detail may not seem like much, but it is one of the earliest references to the concept of magnetic declination. In the West, this observation dates back as early as 1300 AD and compasses were made with an indication of declination in about 1450 AD (Good, 1987). 沈括Shěn Kuò reported this in 梦溪笔谈 *Mèng Xī Bǐ Tán* which was published in 1088 AD. 沈括 Shěn Kuò did not understand why this was the case but he reported it all the same. Sometimes when we do an experiment we do not get exactly what we expected. Those deviations from our theory often wind up pointing to some new phenomena that we were not aware of. We will expand on this seemingly unimportant comment and use it to discover some important physics. The remaining text tells us how to build a compass. We will follow these directions in the experimental section (Chapter 4) when we make simple compasses and use them to measure magnetic fields. The Chinese word for compass is 指南针 zhǐnánzhēn, which more literally means a south-pointing needle. 指 zhǐ has several meanings including toe, finger, indicate, and point. The next character 南 nán means the south or southern. Finally, 针 zhēn means needle, pin, tack, or sometimes it means acupuncture. So when you put it all together you have a south-pointing needle.

While we are on the subject of Chinese characters, let's also learn the characters for the cardinal directions. Figure 1.2 shows a Chinese compass with the characters for north, south, east and west.

Figure 1.2: A Chinese compass showing the characters for the cardinal directions.

At 0° we have the character 北 běi which means north. Opposite that at 180° we have 南 nán for south. Note that the top of the character faces the number. Next, we have 东 dōng at 90° which is east. Remember that in this book we use simplified characters. The character shown on the compass is in traditional form, which is 東 dōng. Finally, at 270° we have 西 xī which means west.

Before we dive into the physics of magnetism it is interesting to look into the etymology of some of the Chinese characters associated with the magnetic compass. Most Chinese characters have a story behind them. Unfortunately, many of these stories have been lost or even contested by scholars. Very ancient forms of a character can look different from the modern form. Sometimes we can see the etymology of the character by examining related pictographs.

Figure 1.3: Ancient forms of Chinese characters for the cardinal directions. From left to right: 北 běi (north), 南 nán (south), 東 dōng (east), 西 xī (west).

The character for north is 北 běi and it appears to be a picture of two people standing back to back. The emperor would sit facing south with his back to the north. Remember the point about facing south to see the Sun? The next character, 南 nán or south, is somewhat difficult to interpret. Some scholars have suggested that it is composed of a plant sprouting from a container. 東 dōng, meaning east, seems to be composed of two parts. A circle with a line or dot in the center is an ancient form of the character for the sun. The vertical line with diagonal lines branching out represent a tree or trees. Thus, some scholars see this as the Sun rising in the east as viewed through the trees. When the Sun sets in the west how do the birds respond? 西 xī, or west, in this representation appears to be a bird that has returned to its nest to roost at sunset. As with most ancient stories, there are other representations of this character all with different interpretations.

Now we can return to the physics of the compass and see what we can learn. Why does the needle orient itself? If you hang the needle from a thread as described and give it a tap you will find that after swinging around for a little bit, the needle returns to its original position. Most people know that if you place two like magnetic poles near each other, then they will repel. Similarly, two unlike poles will attract. By rubbing the needle on the lodestone the needle has become magnetized, with a north-seeking pole at one end and a south-seeking pole at the other end. Notice we used the term "north seeking." That is exactly how we determine which end of a magnet is either north or south. We let the magnet orient and name the pole after the general direction it points. So what attracts the north-seeking pole? To be attracted, the pole exerting the force must be of the opposite magnetic polarity. We usually do not think about this, but that implies there must be a south magnetic pole in the direction of Earth's geographic North Pole. Likewise, the Earth's geographic South Pole must act as if it is the north pole of a magnet. Notice that we are using terms to distinguish the geographic pole from the geomagnetic pole.

1.2   MAGNETIC DECLINATION

What can we learn from the observation that the needle does not point due south? If we represent the Earth as behaving like a giant bar magnet, as shown below in Figure 1.4, we are forced to conclude that the magnetic poles are not aligned with the geographic poles. This misalignment gives rise to a phenomena known as magnetic declination (磁偏角 cípiānjiǎo).

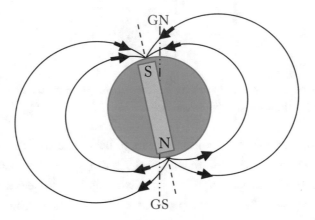

Figure 1.4: Representing the Earth's magnetic field as a simple dipole produced by a bar magnet with magnetic poles (N-S) rotated from the geographic poles (GN-GS).

Now an obvious question arises about how we know the orientation of the geographic north and south poles. 沈括 Shěn Kuò knew how to answer that question very well. He was, at one time in his career, engaged in making astronomical measurements for the purpose of setting the calendar. We take calendars for granted but there was a tremendous amount of calculation and observation involved. If you observe the motion of the stars, the sky seems to rotate about points called the celestial poles. For 沈括 Shěn Kuò the important point would be the North Celestial Pole. This apparent rotation of the sky is of course the result of the Earth's rotation, but from our point of view, it appears that the sky is rotating. Figure 1.5 shows how the sky would appear to rotate over a period of two hours. This image is from the point of view of an observer in Cambridge, UK at about 51° N latitude. The image is a composite of star locations taken every 15 minutes from 8 PM to 10 PM near the end of September. The last image of the sequence is brighter and clearly shows the constellation Usra Major, the Andromeda galaxy (M31), and supernova remnant 1680 (Cassiopeia A). These objects would be smeared out by rotation and we have made them artificially brighter than they would appear to the naked eye just for reference. The field of view is about 44° wide. At the very center of this rotation is the North Celestial Pole, which is indicated by a small cross. Adjacent to the North Celestial Pole is the star Polaris. We can use the Big Dipper asterism, which is part of the constellation Ursa Major, to find the star Polaris. As the tradition goes, you draw a line that connects the stars Merak and Dubhe. Now extrapolate this line until you come to a bright star. That star is Polaris and it is not exactly on the line, but it is close enough to be easily recognized. Since Polaris is near the North Celestial Pole it is often called the North Star. That is fine for us as modern people, but in ancient times Polaris was not that close to the North Celestial Pole.

Figure 1.5: Apparent rotation of the sky over a period of 2 hours. Stars are shown in 15-minute intervals for an observer at about 51° N latitude.

For an observer living at the time of 沈括 Shěn Kuò, Polaris was much further from the North Celestial Pole. Figure 1.6 shows the orientation of Polaris with respect the North Celestial pole as viewed from 镇江 Zhènjiāng, the modern name for the city where 沈括 Shěn Kuò died. In the year 1095, at the time of his death, Polaris and the North Celestial Pole were separated by about 5.7°. In the year 2019, this separation reduced to only about 0.6°.

Using his astronomical training, 沈括 Shěn Kuò clearly knew the direction of true north and could compare that to the orientation of his compass. Since he was living in China, he could not view the South Celestial pole. The south pole would be below the horizon for him. It may be worth pointing out that at his time and in the present era, there is no South Celestial Pole star.

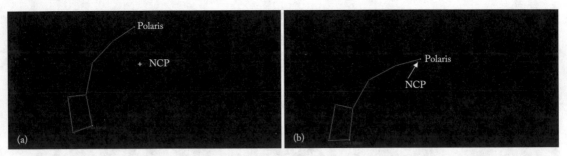

Figure 1.6: Separation between the North Celestial Pole (NCP) and Polaris near midnight around the Autumnal equinox, as viewed from 镇江 Zhènjiāng: (a) in the year 1095 when 沈括 Shěn Kuò died and (b) in the year 2019. Both images are to the same scale.

1.3   MAGNETIC POLES OF THE EARTH

Now let us examine Figure 1.4 a bit more closely. Notice the arrows are pointing away from the north pole of the bar magnet and into the South Pole. This is a general convention when representing any magnetic field. You will also observe that the magnetic field of the Earth is three dimensional. A two-dimensional compass held parallel to the ground can move freely in the horizontal plane. It may tilt slightly up and down, but it is generally limited by the mechanical system required to support the needle. Thus, the compass needle indicates the horizontal component of the Earth's magnetic field. Since the magnetic field is three dimensional, there is also a vertical component. Close examination of Figure 1.4 shows that at the pole the magnetic field has a vertical component that points either up or down and no horizontal component. In reality the pole is not an infinitesimal point but rather distributed over a large area, so the field is not exactly vertical. The poles themselves are not fixed points but can wander over distances of tens or even hundreds of kilometers due to daily variations and magnetic storms. The horizontal component near the poles can be very weak with most of the contribution in the vertical component. If the torque produced by the horizontal component of the field is weak comparted to the frictional torque, the compass needle may not respond and the compass will have erratic behavior.

If we take a compass and turn it on its side it becomes what is known as a dip needle or dip circle. Such a device is shown below in Figure 1.7. Note that the dip angle is measured from the horizontal position. On the magnetic equator, the dip angle is zero. If you move north of the magnetic equator, the north end of the dip needle dips below horizontal. This downward angle is considered to be a positive value. South of the magnetic equator the south end of the needle dips below the horizontal and the angle is considered to be negative. The dip angle shown in Figure 1.7(a) is about +62°. Magnetic dip can be used to find the location of the magnetic pole. In Figure

1.7(b) a small magnet is placed below the needle to simulate the magnetic pole. This causes the needle points straight down with a 90° dip angle.

 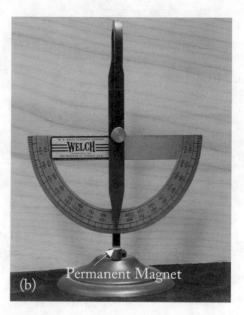

Figure 1.7: A device for measuring the magnetic dip angle. The angle is measured below the horizontal: (a) magnetic dip of local field and (b) needle points to the pole of a magnet placed on the base.

The actual location of Earth's magnetic pole was first found by magnetic dip measurements. Early on people thought that it might be possible to find one's latitude or longitude from measuring the magnetic declination and magnetic dip. In 1600, William Gilbert published his work *De Magnete* in which he described numerous experiments that demonstrated the magnetic properties of the Earth. Using a small lodestone ground into the shape of a ball, he was able to deduce that the magnetic dip angle would change as you moved along the surface of the Earth. He was a great proponent of using magnetic declination and inclination (dip) as a means of determining one's latitude and longitude. As magnetic survey data was amassed, the true picture of the Earth's magnetic field began to appear. Gilbert's theory suffered the same fate as many other theories in the history of science. In reality, the actual data was a lot more complex and that was the end of his idea. The National Oceanic and Atmospheric Administration provides charts and models of the Earth's magnetic field through the National Centers for Environmental Information (NCEI). Charts and mathematical models of the Earth's magnetic field can be freely accessed at https://www.ngdc.noaa.gov/geomag/declination.shtml. These maps do not require any special permissions for use.

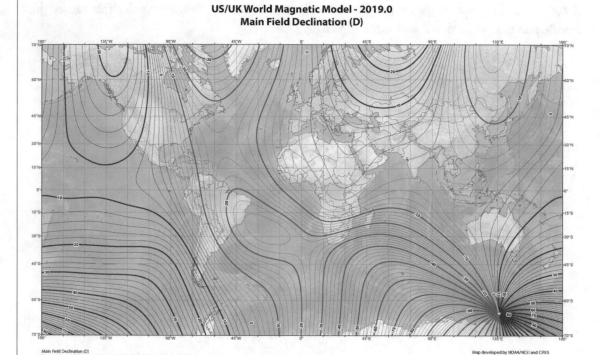

Figure 1.8: Worldwide declination map from the National Centers for Environmental Information (NCEI). Note the convergence of blue and red contours indicating the position of the south magnetic pole.

Figures 1.8 and 1.9 clearly show the difficulty of navigating by magnetic declination or inclination. You might naively think that by following a line of constant declination you would also be on a line of constant longitude. The wide bends and twists could put you hundreds of miles off course. The line of zero inclination (dip) does not follow the equator but wanders above and below it. In the vicinity of the poles the situation is far worse. To make matters even more complex, the Earth's field changes with time. Few people realize that the magnetic north pole has wandered considerably over the past few hundred years. This phenomenon is clearly shown is Figures 1.10 and 1.11.

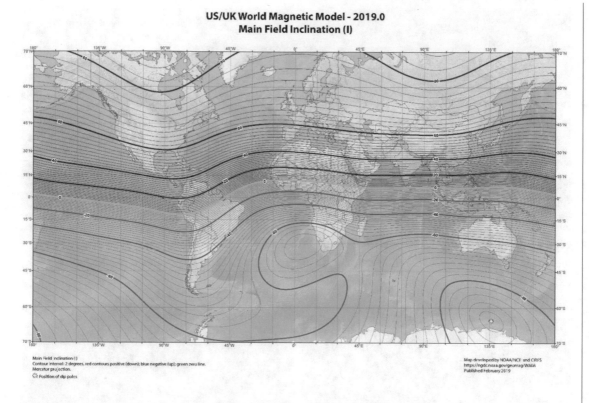

Figure 1.9: Worldwide magnetic inclination (dip) from the National Centers for Environmental Information (NCEI). Note the position of the dip pole with an inclination angle of -90°.

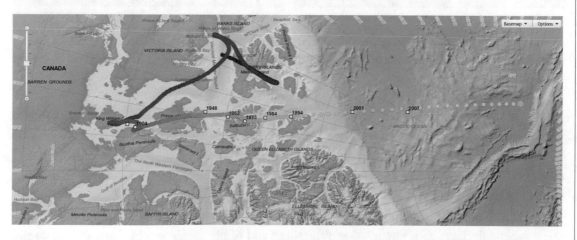

Figure 1. 10: Wandering of the North Magnetic Pole. Image from the National Centers for Environmental Information (NCEI). Yellow squares are observed data; circles are modeled results dating back to 1590 (blue).

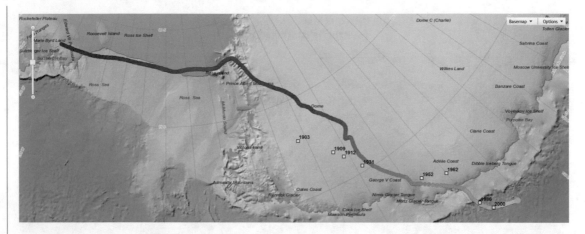

Figure 1.11: Wandering of the South Magnetic Pole. Image from the National Centers for Environmental Information (NCEI). Yellow squares are observed data; circles are modeled results dating back to 1590 (blue).

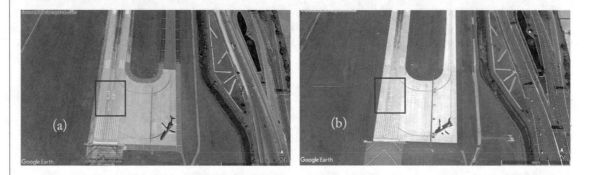

Figure 1.12: Airport runway designations are changed in response to Earth's changing magnetic field. This runway at Tampa International Airport was previous designated 36L and in 2012 it was changed to 1 L.

Figure 1.12(b) shows a runway number of 1L, but this is exactly the same runway as shown in Figure 1.12(a). The difference is that the current (2019) magnetic heading for the runway is 7°. In assigning a runway number you round to the nearest 10° and then drop the last digit. So the 7° rounds up to 10° which then becomes 1. The opposite end of the runway has a magnetic heading of 187° so it is designated as 19.

If 沈括 Shěn Kuò could somehow perform the same experiment in 2019 he would find that the magnetic declination is about -6° and the south end of his needle would point slightly to the east, as shown in Figure 1.13. However, if he tried it in 1628, the International Geomagnetic Reference Field (IGRF2015) model predicts a barely noticeable angle of about 0.2° to the southwest. Keep in mind that it is only a model and the real situation is not guaranteed to be accurate.

Figure 1.13: **Near Meng Xi Park in 镇江市 Zhènjiāngshì. Blue line shows Magnetic North-South** with declination of about -6° in the year 2019.

1.4   REPRESENTING THE MAGNETIC FIELD OF THE EARTH AS A VECTOR

We can represent the magnetic field of the Earth at any point as a vector B, as shown in Figure 1.14. The components of the vector are the north intensity (x), the east intensity (y), and the vertical intensity B_z. The horizontal intensity, $\mathbf{B_h}$, is a vector that lies in the x-y plane and points in the direction of a properly balanced compass needle. Normally, a compass needle has its center of mass at a slight distance away from the support to counteract the dip and keep the needle nearly horizontal. There are two important angles D and I which are measured in degrees. Angle D is the magnetic declination which is measured with respect to true north. A positive angle indicates that magnetic north is east of true north and a negative value indicates magnetic north is west of true north. Angle I is called the inclination and is measured with respect to the horizontal plane. If the magnetic field vector points down toward the surface of the Earth then the angle is considered positive. If the field vector is above the horizontal plane, then the angle is considered negative.

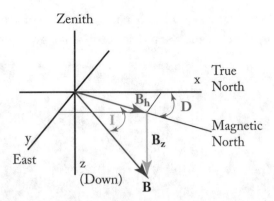

Figure 1.14: Representation of the magnetic field vector in terms of the magnetic elements.

Magnetic fields are normally measured in units of Tesla (T) but this is a huge unit compared to the small fields generated by the Earth. For geomagnetic fields we often use units of nano-Tesla ($1\ nT = 1\times10^{-9}\ T$) or Gauss (G). Small magnetic fields are more conveniently expressed in Gauss since $1\ T = 1\times10^4$ G. The magnetic field of the Earth is typically in the range of about 0.25–0.65 G. We covered the basic concepts of vectors in *Teaching Physics through Ancient Chinese Science and Technology*, Chapter 3. Students should review that section as we are now going to perform some calculations based on the magnetic field of the Earth.

The National Center for Environmental Information has an online magnetic field calculator that can be accessed at: https://www.ngdc.noaa.gov/geomag/calculators/magcalc.shtml#igrfwmm. The magnetic field calculator uses the World Magnetic Model to predict the magnetic field at any position of the Earth. 沈括 Shěn Kuò's former residence is located in 镇江市 Zhènjiāngshì. The position of 镇江市 Zhènjiāngshì is Latitude 32° 12' 39"N and 119° 27' 18" E. Using this position the magnetic field calculator gives us the following results for the year 2019.

Table 1.1: Magnetic Field at 镇江市 Zhènjiāngshì from World Magnetic Model		
Declination D (°)	**Inclination I (°)**	**Total Intensity B (nT)**
-6.044	48.90	4.998×10^4

The declination angle (D) is negative which tells us that magnetic north is west of true north by about 6°. This implies that the south pointing end of the compass needle would point slightly to the east as it did in the days of 沈括 Shěn Kuò. We also see that the inclination (I) is positive so that compass needle would dip down. From this information we can calculate the horizontal magnetic field B_h and the components along the axes. Figure 1.15 shows the orientation of $\mathbf{B_h}$, which lies in the horizontal x-y plane. The total field $\mathbf{B}$ is 48.90° below the plane. From the figure we can see that the field $\mathbf{B}$ is the hypotenuse of a triangle and that B_h is the side of the triangle adjacent to the angle I (48.90°). The equation for the horizontal component is then

$$B_h = B\ Cos\ (I) = 4.998x10^4\ nT\ x\ Cos\ (48.90°) = 3.286x10^4\ nT$$

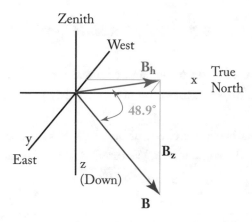

Figure 1.15: Orientation of the horizontal field $\mathbf{B_h}$ and the total field $\mathbf{B}$.

From Figure 1.15 we also see that the z-component B_z is the side of the triangle that is opposite the inclination angle. This means that the component B_z is given by

$$B_z = B\ Sin\ (I) = 4.998 \times 10^4\ nT \times Sin\ (48.90°) = 3.766 \times 10^4\ nT$$

Now that we know the magnitude of $\mathbf{B_h}$ and its direction with respect to the coordinate axes we can calculate the x and y components of the vector. The x-component tells us the strength of the field in along the true north-south axis and the y-component tells us the strength in the along the east-west axis.

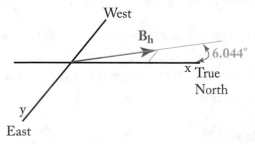

Figure 1.16: Orientation of the horizontal field $\mathbf{B_h}$ looking down on the x-y plane.

The horizontal field $\mathbf{B_h}$ is the hypotenuse of the triangle and the side parallel to the y-axis is opposite of the declination angle D. The side of the triangle that is adjacent to the angle is along the x-axis. Thus, the field components are given by

$$B_x = B_h \, Cos\,(D) = 3.286 \times 10^4 \; nT \times Cos\,(6.044^o) = 3.268 \times 10^4 \; nT \quad,$$

and

$$B_y = -B_h \, Sin\,(D) = -3.286 \times 10^4 \; nT \times Sin\,(6.044^o) = -3.460 \times 10^3 \; nT \quad.$$

Notice that we could have used $D = -6.044^o$ and obtained the same result. Here we inserted the negative sign into the equation, since the component obviously points in the negative y-direction, or west.

We can perform a mathematical check of our work be recalculating the magnitude of the total magnetic field **B** using the components. We can find the magnitude from the equation

$$B = \sqrt{B_x^2 + B_y^2 + B_z^2}$$

$$= \sqrt{\left(3.268 \times 10^4 \; nT\right)^2 + \left(-3.460 \times 10^3 \; nT\right)^2 + \left(3.766 \times 10^4 \; nT\right)^2} = 4.998 \times 10^4 \; nT.$$

As expected, this is the same as the total magnetic field magnitude that we started with. Always look for a way to mathematically check your work.

Expressing the magnetic field as $4.998 \times 10^4 \; nT$ seems a bit cumbersome and if we want to make comparisons to other sources of magnetic fields we should express the answer in one of the more commonly used units. We can convert from nT to T using $1nT = 1 \times 10^{-9}T$. In the field of geomagnetism it is common to measure the Earth's magnetic field in units of microtesla, where $1\mu T = 1 \times 10^{-6} \; T$. Converting to these other units we have then have

$$4.998 \times 10^4 \; nT \times \frac{1 \times 10^{-9} \; T}{1 \; nT} = 4.998 \times 10^{-5} \; T = 4.998 \times 10^{-5} \; T \times \frac{1 \times 10^6 \; \mu T}{1 \; T} = 4.998 \times 10^1 \; \mu T \quad.$$

A $1T$ magnetic field is very strong. Just for comparison, those common refrigerator magnets we use to hang up notes are several mT ($1mT = 1 \times 10^{-3}T$). Clinical Magnetic Resonance Imaging (MRI) scanners are tens of thousands times stronger than the Earth's field. They usually operate in the several Tesla range.

1.5   THE 罗盘 LUÓPÁN

One might think that the original use or the compass in Ancient China was for navigation. Actually, it was first used as a 气 qì meter. At least, that is what the geomancers believed they were measuring. The compass they used for this work is called a 罗盘 luópán. Figure 1.17 shows the needle of a 罗盘 luópán reacting to a lodestone or 慈石 císhí.

Figure 1.17: A Chinese 罗盘 luópán reacting to the magnetic field of magnetite.

The ancient Chinese devised a complex system called 风水 fēngshuǐ as a way of describing what they thought of as the flow and interaction of 气 qì. This system is reflected in the bewildering series of rings and indicators found on a 罗盘 luópán. It is not the purpose of this text to go into the complexities of their ideas, but we can try to understand it on some very basic level. 风水 fēngshuǐ, or geomancy as it is known in the West, literally means wind (风 fēng) and water (水 shuǐ). Proponents of this philosophy believe that 气 qì is a form of energy that, like water, flows through our surroundings. It rides upon water and pools where water pools or flows where water flows. In their understanding of the natural world, they believe that it has an effect on buildings, tombs, and their residents. These effects manifest themselves in areas of fertility, luck, wealth and the general happiness of individuals. The ebb and flow 气 qì is seen as an environmental factor that affects mankind's overall health and happiness. A 罗盘 luópán, along with other geological features, is seen as a way of diagnosing or plotting that flow. 风水 fēngshuǐ ideas are often reflected in the architecture of buildings. The concept still survives to this very day and reflected in modern Asian architecture. Modern adepts make a good living advising their clients about the placement and orientation of buildings.

In our discussion up to this point we have only dealt with naturally occurring magnetic fields. Much of our modern technological society owes its very existence to the science of electromagnetism. Our ability to create and control magnetic fields is of very recent origin dating back to roughly the early 19th century. As we shall see the, magnetic compass helped to usher in a new technological age. Francis Bacon may have only thought of magnetism in terms of navigation, but about 200 years after his death and new use for magnetism was discovered.

In *Teaching Physics through Ancient Chinese Science and Technology*, we tried to show that early science in the West was not always described in a mathematical formalism. Two of the greatest early experimenters in the field of electromagnetism were Michael Faraday and Benjamin Franklin. Yet, they did not express their ideas in a mathematical way. Their meticulous experimentation often lead them through a convoluted maze of observations and dead ends. Ultimately, their observations were coupled to mathematics by great theorists such as Maxwell and a more coherent picture emerged. 沈括 Shěn Kuò was, in fact, educated in many practical applications of geometry and trigonometry. Modern scholars often criticize him for not developing an integrated framework for his diverse knowledge base. Why should he? He did have a framework that involved things like 气 qì and 五行 wǔxíng.

1.6   OBSERVATIONS OF MAGNETISM BY FRANKLIN, FARADAY, AND ØRSTED

1.6.1   FRANKLIN'S EARLY EXPERIMENTS

Before we look at the experiments that connected electricity and magnetism, we should survey some of the ideas and technology that were being circulated at the time. Physics did not develop in a vacuum and there were technological and philosophical forces driving many discoveries. We know that Franklin connected lightning and electric charge but there is also an interesting experiment he conducted that may have inspired the ultimate connection between electric currents and magnetism. The following is an excerpt taken from Franklin's correspondence (Franklin, 1769).

> June 29, 1751. In Capt. Waddell's account of the effects of lightning on his ship, I could not but take notice of the large comazants (as he calls them) that settled on the spintles at the top-mast heads, and burnt like very large torches (before the stroke)… His compasses lost the virtue of the load-stone, or the poles were reversed; the North point turning to the South. By Electricity we have (here at Philadelphia) frequently given polarity to needles, and reversed it at pleasure. Mr Wilson, at London, tried it on too large masses, and with too small force. A shock from four large glass jars, sent through a fine sewing needle, gives it polarity, and it will traverse when laid on water.—If the needle when struck lies East and West, the end entered by the electric blast points North. —If it lies North and South, the end that lay towards the North will continue to point North when placed on water, whether the fire entered at that end, or at the contrary end. The Polarity given is strongest when the Needle is struck lying North and South, weakest when lying East and West; perhaps if the force was still greater, the South end, enter'd by the fire, (when the needle lies North and South) might become the North, otherwise it puzzles us to account for the inverting

of compasses by lightning; since their needles must always be found in that situation, and by our little Experiments, whether the blast entered the North and went out at the South end of the needle, or the contrary, still the end that lay to the North should continue to point North. In these experiments the ends of the needles are sometimes finely blued like a watch-spring by the electric flame. This colour given by the flash from two jars only, will wipe off, but four jars fix it, and frequently melt the needles.

This experiment suggests that somehow the "electric fire" or current flowing through a needle can change the magnetic orientation of the needle. Franklin seems to be suggesting that it is the electric current which is the source of the magnetization. Notice that Franklin points out the needle became blued. Steel becomes blued when heated due to an oxide layer that forms on the surface. We know that the needles can even be melted by the "electrical fire" that passes through them. Little did Franklin know that he just stumbled across a secret piece of Chinese military technology. From about the time of the 宋朝 Sòngcháo or Song Dynasty this very same effect was known and considered classified technology. The ancient Chinese knew that if you heated an iron needle and then cooled it quickly, with the correct orientation to the North-South axis, you could make a compass needle without rubbing it against 慈石 císhí. The effect is known as thermoremanent magnetization and is also seen in igneous rocks. When the needle is red hot it acquires its magnetization from the magnetic field of the Earth. The magnetic orientation is then "frozen in" when the iron is cooled. When this effect occurs in igneous rocks it records the local properties of the magnetic field in the rock. Thus, we have a historic record of Earth's past magnetic field alignment stored in rock. The study of this historical record is called Paleomagnetism. Paleomagnetic studies reveal that the Earth's magnetic field has reversed polarity some 183 times over the last 8.3×10^7 years, with the last event about 7.8×10^5 years ago.

At the time of Franklin there were no chemical storage batteries as we understand them. Franklin did coin the word "battery" but for him it referred to a collection of Leyden jars used to store charge arranged like a battery of cannon. Thus, it was difficult to provide a continuous charge through a wire. His Leyden jars were capacitors that stored charge and then delivered that charge very quickly through a spark discharge. That is what he means when he uses terms such as "jars," "blast," and "flash."

Franklin's ideas about the connection between electricity and magnetism evolved in time and by 1773 we find him recanting some of his ideas and thinking about magnetism as some sort of magnetic fluid that flows inside all iron-bearing materials (Franklin, 1970).

As to the magnetism, which seems produced by electricity, my real opinion is, that these two powers of nature have no affinity with each other, and that the apparent production of magnetism is purely accidental. The matter may be explained thus.

1st The Earth is a great magnet.

2dly There is a subtle fluid, called the magnetic fluid, which exists in a ferruginous bodies, equally attracted by all their parts, and equally diffused through their whole substance; at least where the equilibrium is not disturbed by a power superior to the attraction of the iron....

9thly An electric shock passing through a needle in a like position, and dilating it for an instant, renders it, for the same reason, a permanent magnet; that is, not by imparting magnetism to it, but by allowing its proper magnetic fluid to put itself in motion.

Science is often taught in a way that gives the impression that scientific discoveries just fell out of the sky in their full and complete form. Students are usually presented with full and complete theoretical systems. Very few science teachers show incomplete models or lead their students through the evolution of scientific understanding. They just present the end product and require memorization of related facts. There was a time when Newton was just Isaac and he had detractors who railed against his ideas. The cause of magnetism created a great deal of debate in the scientific community and various theories were proposed. As is often the case in science, better experimental equipment lead to deeper experimentation. This was clearly the case with the invention of a continuous source of electric current. The production of a continuous current by chemical means did not come about until 1799 when Alessandro Giuseppe Antonio Anastasio Volta invented the electric column or what we now call the voltaic pile. Yes, the unit of electric potential sounds much like his name and that is where we get the term Volt. Volta is remembered for quite a few interesting discoveries and inventions one of which came as a direct result of reading the work of Ben Franklin. In 1778 he isolated the chemical compound Methane (CH_4). Franklin had written a paper describing what he called "flammable air" and in 1776 Volta found such "flammable air" in Lake Maggiore. With Volta's new chemical battery, experimenters were able to provide a continuous electric current and that simple step lead to a wide range of electromagnetic discoveries.

1.6.2   ØRSTED'S EXPERIMENTS

Some 50 years after Franklin's musing about the connection between electricity and magnetism the decisive experiment was performed. Hans Christian Ørsted, Danish physicist and chemist, demonstrated that a compass needle can be deflected when electric current from a battery passed through a wire adjacent to the compass. Ørsted is pronounced as UR-sted and is often rendered as Oersted in English. There are many different stories about how this experiment came about and whether or not it was an accidental discovery but we will not go into all of that here. What we do know is that Ørsted demonstrated the effect on April 21, 1820 during a lecture. Perhaps Ørsted anticipated the outcome of this experiment, as he was philosophically predisposed to the idea that electricity, magnetism, heat, and light were all interconnected. Ørsted's Ph.D. dissertation, *The Architectonics of Natural Metaphysics,* was an examination of ideas based on the work of the Prussian German philosopher Immanuel Kant. Kant's ideas on the unity of nature and relationships between natural phenomena seemed to have struck a chord with Ørsted and his close friend Johann Wilheim Ritter who, among other things, discovered ultraviolet radiation. William Herschel discovered the "heat rays" of infrared radiation in 1800 and Ritter's ideas about symmetry between the forces of nature spurred him to look for "cooling" rays at the opposite end of the visible spectrum. Instead, he found the very energetic rays we now call ultraviolet.

This little journey through history and philosophy may seem like a bit of a distraction if not understood properly. Modern day scientists often dismiss the contributions of ancient China as not being truly "scientific" and full of unproven metaphysical ideas. Of course, the idea of "scientific" or a "scientist" did not even exist in those ancient times. They did have their own understanding of how the world worked and their place in the cosmos. There seem to be interesting parallels between the systems of ancient China and the state of experimental sciences of the 18th and 19th century. The stereotype of the cold detached scientist who evaluates everything by the scientific method is not so easy to find in the history of science. Ørsted's experiments provided the data that French mathematician and physicist, André-Marie Ampère, used to create his more mathematical theories of electromagnetism. This unity between the physical world and mathematics not only provides us with a way of explaining what we observe but gives us a predictive power that has become the hallmark of the modern scientific method.

We will now examine the magnetic field created by a wire wound into several different geometric shapes. The equation for the magnetic field of a long straight wire is the subject of one of the laboratory experiments described in Chapter 4 of this work. Ørsted found that a compass needle was deflected by the presence of a current in a wire adjacent to a compass. This could only be the case if the current produced a magnetic field. More than that, he found some curious facts about how the compass needle deflection was related to the orientation of the wire and the direction of the current through the wire. In the following figures we illustrate some of the facts that he discovered and the conclusions one can draw from them. Let us first examine a long straight wire oriented above a compass, as shown in Figure 1.18. Notice that the wire is located directly above the compass and runs parallel to the needle. The wire is actually a brass rod capable of carrying a very high current without getting hot. This is especially important for the sake of the experimenter's safety and that of the plastic compass housing. The rigidity of a rod, compared to a flexible wire, helps to maintain a fixed orientation between the current and the magnetic compass.

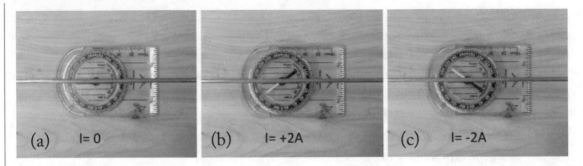

(a)   I = 0   (b)   I = +2A   (c)   I = -2A

Figure 1.18: A compass needle deflected by various electrical currents passing through a long straight metal rod.

With no current through the rod (I=0), the needle is aligned along the horizontal component of Earth's magnetic field. When current is allowed to flow through the rod, as shown in image (b), the compass needle is deflected. The north-seeking pole of the compass needle is deflected away from the rod. Recall Figure 1.4 which shows that the magnetic pole near the Earth's geographic north pole is actually the south pole of the Earth's magnetic field. The north-seeking end of the compass needle must actually be the north magnetic pole of the compass needle. That is the red colored end of the compass needle.

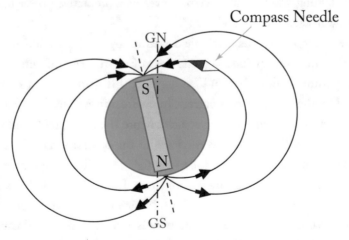

Figure 1.19: The red or north-seeking end of the compass needle is attracted the south pole the Earth's internal magnet.

Now that we know the magnetic polarity of the compass needle, what does that imply about the magnetic field produced by the rod? In the plane of the compass needle, which is below the rod, the red end must experience a magnetic field that behaves like the north pole of a magnet. Notice that when the current direction is reversed (I= −2A), the red end is deflected away from the rod but on the opposite side. This implies that the magnetic field of the rod is now pointing in the opposite direction where it intersects the plane of the needle.

Now let us examine what happens when we move the compass above the rod. Figure 1.20 shows the compass deflection for two different cases. The current magnitude and direction are the same for each case but the position of the compass is now either above or below the rod. The distance between the rod and the compass needle is about the same and no attempt was made to control that parameter when photographed.

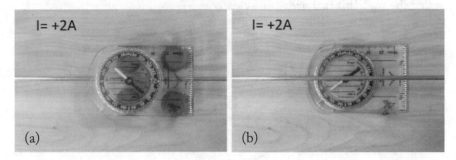

Figure 1.20: The compass is positioned above and below the rod with the same current magnitude and direction.

This result shows that the magnetic field above and below the rod does not point in the same direction. The direction of the field above the rod is opposite of the direction below the rod. How can that be? One way to understand this is to think of the magnetic field as circulating around the rod. This circulation is demonstrated by several compasses arranged around the rod, as shown in Figure 1.21.

Figure 1.21: The magnetic field circulates around the wire as shown by the compass needles.

Each compass needle aligns itself with the horizontal component of the magnetic field. With only six compasses we obtain a polygon. If we were to increase the number of compass needles, we would see our polygon describe something that looked more like a circle. Notice the direction of the compass needles depends on the direction of the current. With no current, the compasses naturally point north and there is a slight interaction between each compass. When a current of 10 Amps passes through the rod, with the current directed downward, the compasses point along a clockwise circulation. The direction of the circulation reverses when the current is directed up the rod. This idea of the magnetic field circulating around the rod explains why the compass deflection changes above and below the rod. This is shown more clearly in Figure 1.22.

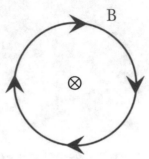

Figure 1.22: Magnetic field produced by a long straight rod. Current directed into the page.

As Figure 1.22 demonstrates, the magnetic field B at the top of the circle points in the opposite direction of the magnetic field at the bottom of the circle. Here the current is directed into the page. For a quantity pointing into the page we use an x and for one pointing out of the page we use a dot. These are somewhat reminiscent of a vector arrow coming toward you with the dot representing the tip of the arrow. This convention is shown below in Figure 1.23.

In   Out

Figure 1.23: Representation of a vector pointing into the page and out of the page.

We can predict the direction of the magnetic field by using the "right-hand rule." The basic idea of the right-hand rule is that you grasp the conductor in your right hand with your extended thumb pointing in the direction of the current. Your fingers then curl around the wire in the direction of the magnetic field. Applying the right-hand rule to our rod gives the result shown in Figure 1.24.

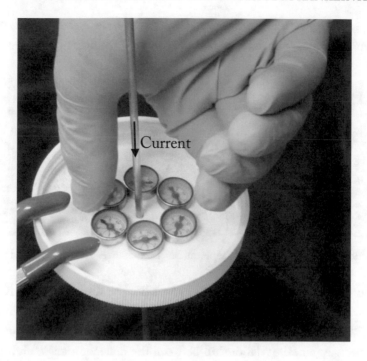

Figure 1.24: Applying the right-hand rule. Thumb in the direction of the current, fingers curl in the direction of the magnetic field.

Now that we know how to predict the direction of the magnetic field we should consider its strength or magnitude. Figure 1.25 shows how the deflection of the compass needle depends on the strength of the current through the rod.

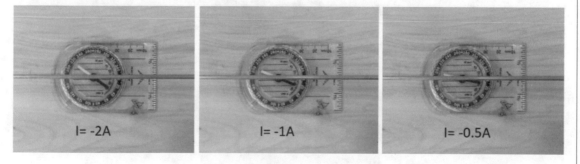

Figure 1.25: The deflection of the needle depends on the current through the rod.

The deflection of the compass needle gives an indication of the magnetic field strength. It is true that decreasing the current decreases the magnetic field strength. The relationship, however, is not a linear one. Half the current does not give half the angle. We will explore this relationship in the laboratory experiment described in Chapter 4, see Figure 4.14. Not only does the magnitude of the current influence the magnetic field strength, but so does the distance between the rod and the compass. We can see this behavior demonstrated in Figure 1.26.

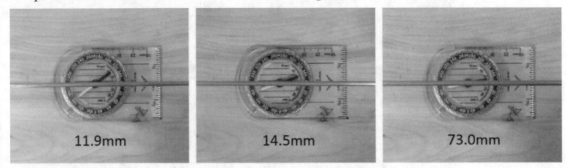

Figure 1.26: The deflection of the compass depends on the distance between the rod and the compass.

In the series of photographs shown in Figure 1.26, the distance between the rod and the compass increases. At is closest distance (11.9 mm) the deflection is greatest indicating that the field is strongest. We can barely see the deflection when the rod is positioned about 73 mm from the plane of the compass. Again, the relationship between the deflection angle and the field strength is not simply linear.

In the experiments we have described thus far, the rod has been oriented parallel to the initial position of the compass needle. What if the rod is oriented perpendicular to the needle? The result is shown below in Figure 1.27.

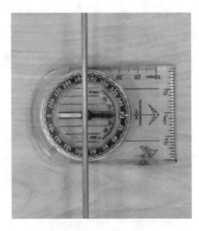

Figure 1.27: Rod oriented perpendicular to the needle with current of 2.0 A.

In this configuration the magnetic field intersects the plane of the needle parallel to the needle's axis. Thus, there is no torque acting to twist the needle from its initial position.

Magnetic Field Produced by a Long Straight Wire

How does the strength of the field depend on current and distance? The equation for the strength of magnetic field B, produced by a long straight wire carrying a current i, is given by

$$B = \frac{\mu_0 i}{2\pi R},$$

where μ_0 is a constant is called the permeability of free space and is defined as exactly

$$\mu_0 = 4\pi \times 10^{-7} \; Tm \; A^{-1}.$$

In the equation for the magnetic field, R is the radial distance away from the center of the wire to a point in space. Notice that the units of μ_0 imply that the strength of the field is measured in units of Tesla, and the current is measured in units of Amperes. The direction of the magnetic field is given by the right-hand rule.

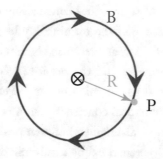

Figure 1.28: Magnetic field at a point P produced by a current flowing in a long straight conductor. Current directed into the page.

The equation for the magnetic field

$$B = \frac{\mu_0 i}{2\pi R}$$

predicts that the magnetic field strength will be proportional to the magnitude of the current that produces it and inversely proportional to the distance from the source. In Chapter 4 of the experimental section, we describe an experiment designed to explore these relationships.

For the situation shown in Figure 1.26, the rod was about 11.9 mm above the compass needle and the current was 2.00A. The magnetic field strength at the compass needle is given by

$$B = \frac{4\pi \times 10^{-7} \; Tm \; A^{-1} \times 2.00 \; A}{2\pi \times 11.9 \times 10^{-3} \; m} = 3.36 \times 10^{-5} \; T = 3.36 \times 10^{4} \; nT.$$

We can compare this value to the strength of the Earth's field given in Table 1.1. Of course, the data in Table 1.1 is for 镇江市 Zhènjiāngshì in China and the photograph was taken in Macon, Georgia where the Earth's magnetic field is different. The ratio of the magnetic field produced by the rod to the magnetic field in Table 1.1 is

$$\frac{3.36 \times 10^4 \, nT}{4.998 \times 10^4 \, nT} = 6.72 \times 10^{-1}.$$

Thus, the field produced by the current in the rod is a little over 2/3 of the field produced by the Earth. The point of this calculation is to show that the magnetic field created by the current is comparable to the strength of the Earth's field.

Shortly after Ørsted's discovery, he wrote a short treatise on his discoveries entitled *Experimenta Circa Effectum Conflictus Electrici in Acum Magneticm*, or in English, *Experiments on the Effects of a Current of Electricity on the Magnetic Needle*. The original Latin edition was transmitted throughout the world and immediately other experimenters reproduced the effects described. This was a great moment in the history of science and conclusively demonstrated the relationship between electricity and magnetism. Some of the terminology he used may seem strange to us. For example, he refers to the magnetic field as an electrical conflict or "conflictus electrici" as if it were a conflict between the positive electricity and the negative electricity flowing in the circuit. The concept of a field was not in general use at this time and even when mechanism of fields were invoked they were more of a mathematical construct than a real phenomenon. There was a time when early experimenters did seem to distinguish between positive and negative charge as if they were two different things. Electricity generated by a galvanic electrochemical cell (a battery) was also suspected of somehow being different from static electricity produced rubbing. The idea of connecting the positive and negative terminals of a battery to produce an electric current was not at all obvious. Some experimenters thought that a battery in an open circuit configuration might produce magnetic effects. Ørsted pointed out these ideas and some other amazing properties of the magnetic field which we reproduce below (Ørsted, 1820).

> It seemed demonstrated by these experiments that the magnetic needle was moved from its position by the galvanic apparatus, but that the galvanic circle must be complete, and not open, which last method was tried in vain some years ago by very celebrated philosophers.... The nature of the metal does not alter the effect, but merely the quantity. Wires of platinum, gold, silver, brass, iron, ribbons of lead and tin, a mass of mercury, were employed with equal success.... The effect of the uniting wire passes to the needle through glass, metals, wood, water, resin, stoneware, stones.... It is needless to observe that the transmission of effects through all these matters has never before been observed in electricity and galvanism. The effects therefore, which

take place in the conflict of electricity are very different from the effects of either of the electricities.

Further on in his treatise Ørsted tells us how the magnetic field interacts with magnetic and non-magnetic materials.

> A brass needle, suspended like a magnetic needle, is not moved by the effect of the uniting wire. Likewise needles of glass and gum lac remain unacted on…. The electric conflict acts only on the magnetic particles of matter. All non-magnetic bodies appear penetrable by the electric conflict, while magnetic bodies, or rather their magnetic particles, resist the passage of this conflict. Hence they can be moved by the impetus of the contending powers.

Magnetic Field Produced by a Solenoid

If we used a wire instead of a rod, we could wrap the wire around a tube and make a long coil. This configuration is called a solenoid. The term solenoid comes from the Greek word σωλήν (sōlḗn), which means a channel, gutter, pipe. A tightly wound coil does have the appearance of a pipe. Figure 1.29 shows the magnetic field configuration of a solenoid. In an ideal solenoid the coils have very little space between them and there is almost no external magnetic field except near the open ends. The magnetic field inside the ideal solenoid is uniform in intensity.

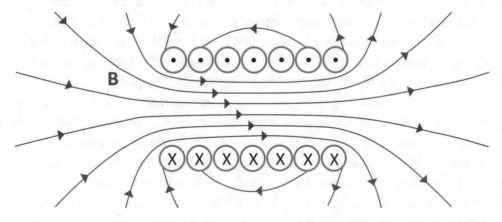

Figure 1.29: Magnetic field of a solenoid.

The geometry depicted in Figure 1.29 is that of a solenoid sliced along a plane parallel to its long axis. The circles with dots and x's represent current flowing out of and into the page, respectively. For a solenoid the net magnetic field is the vector sum of the field produced by each wire. This gives rise to a multiplicative effect related to the number of turns of wire. Deep inside the solenoid, away from the ends, the magnetic field is more uniform and the strength is given by

$$B = \mu_0 n i \ ,$$

where n is the number of turns/unit length. Increasing the number of turns in a given length increases the strength of the magnetic field produced for a given current. However, as we increase the number of turns in a fixed length, the wire would become thinner and that increases its electrical resistance. This means that a higher voltage must be applied for an equivalent current to flow. Suppose we have a solenoid that is wrapped with 20.0 turns of wire per centimeter and it carries a current of 2.00 A. The magnetic field strength is

$$B = \mu_0 n i = 4\pi \times 10^{-7} \ Tm \ A^{-1} \times 2.00 \times 10^3 m^{-1} \times 2.00 A = 5.03 \times 10^{-3} \ T \ .$$

Notice that we converted 20.0 turns cm^{-1} to 2.00×10^3 turns m^{-1} and dropped the "turns" since that is a unitless number. How does this compare to the magnetic field of the Earth given in Table 1.1? The ratio would be

$$\frac{5.03 \times 10^6 \ nT}{4.998 \times 10^4 \ nT} = 1.01 \times 10^2 \ ,$$

or about 100 times the Earth's field shown in Table 1.1. In the previous example using just a single rod the magnetic field strength was only about 2/3 of the Earth's field at the position of the compass. The key idea here is the multiplicative effect of the adjacent wires. The uniform field produced inside the solenoid has many uses for scientific experimentation. Solenoids and the related technology of electromagnets are ubiquitous. In 1824, William Sturgeon demonstrated an electromagnet that could lift about 9 pounds or about 20 times its own weight (Sturgeon et al., 1824). This set off an electromagnet technology race with devices that were eventually able to lift thousands of pounds. Sturgeon started out his career as an apprentice shoemaker turned inventor and physicist. He made the remarkable discovery that by inserting soft iron inside the tube of the solenoid, he could increase the strength of his electromagnet. This was a huge discovery and started a branch of physical research into the magnetic properties of matter. Soft iron does not mean that it is physically soft; it is still a hard material. The softness is in the magnetic response and the fact that the iron loses its magnetic properties when the current is switched off. Along the way he developed a device that produced rotary motion by magnetic fields, the forerunner of the electric motor. Building on the work of Sturgeon, Joseph Henry, the famous American physicist, inventor, and first secretary of the Smithsonian Institution conducted numerous experiments on magnetic devices and made discoveries related to electromagnetic induction. During the years 1831–1832, while at Albany Academy, a college preparatory school for boys, Henry demonstrated an ingenious little device that changed the very nature of how the world communicated. He built a simple device that consisted of a horseshoe-shaped soft iron core wrapped very tightly with insulated wire. The combination of the huge number of turns and the magnetic properties of the soft iron enabled Henry to produce mechanical action via a current transmitted through a very long wire. The electromagnet caused a long rod of permanently magnetized steel to rotate on a pivot and strike an adjacent bell, as shown in Figure 1.30. This is not just the invention of the doorbell, but rather the basic operating principle of the electric telegraph. Henry and Samuel Morse engaged in a long legal battle of the invention of the telegraph. In our modern world, we take for granted the seemingly instantaneous methods of communication and access to information and overlook the impact of the telegraph.

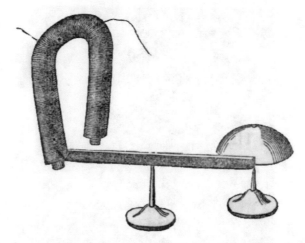

Figure 1.30: Henry's prototype of the electric telegraph. From *Annual Report of the Board of Regents of the Smithsonian Institution showing the Operations, Expenditures and Condition of the Institution for the Year 1857*, p. 105.

Have you used or interacted with a solenoid today? Did you turn a key to start a vehicle? Listen to music projected from a speaker or dynamic headphones? Perhaps you got a drink from a machine that uses an electronically controlled valve such as a beverage dispenser or your refrigerator. Solenoids are used in all of these devices and many more. Current flowing through a coil produces a magnetic field, which in turn, exerts a force on a permanent magnet or magnetic material. This force then results in some form of motion. You may even find yourself inserted into the giant solenoid of an MRI machine, but let's hope not.

1.6.3   FARADAY'S EXPERIMENTS

If a current can produce a magnetic field, can a magnetic field produce a current? If you had this thought, welcome to the world of Michael Faraday. Faraday did ask himself this same question. If you want to answer that question, you need a device to detect electric current. Back in Faraday's time there were not a lot of options for detecting electrical currents. The sensitive meters we have now simply did not exist. You could use your tongue, a fine wire that glows, electroplating, discharge through charcoal, or perhaps a Schweigger multiplier. In 1820, Johann Salomo Christoph Schweigger built a device for detecting electric currents modeled after Ørsted's experiment and harnessing the multiplicative effect of a coil. Instead of just one wire passing above a compass, Schweigger found that he could increase the deflection angle by wrapping multiple turns of wire around the compass. This multiplicative effect is much the same as the solenoid. We illustrate the "multiplier" in Figure 1.31 shown below.

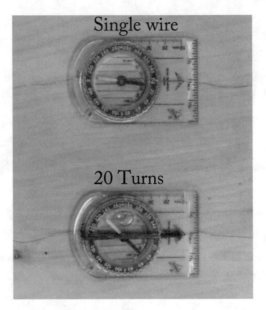

Figure 1.31: The Schweigger multiplier or Galvanometer. The current through each wire is the same but the lower coil consists of 20 turns while the upper is just a single wire.

The multiplier is also known as a Galvanometer, which is named after Luigi Galvani, who in 1780 discovered that an electric spark made the legs of a dead frog twitch. In Figure 1.31, the upper compass has only a single wire passing over it, while the lower compass has 20 turns of wire wrapped around it. The current through the wires are exactly the same. The lower compass can detect a much smaller current. The illustration is slightly misleading since the angle is not linearly related to the current through the coil, but overall it is a more sensitive device. This was the leading technology of the time and Faraday had a very sensitive version that he used to detect small electrical currents. New or improved measurement technologies often lead to new discoveries in physics. This has been the way since we started making experimental apparatus and is an important lesson to keep in mind.

Induction by Time Varying Current

Returning to Faraday, in 1831 he gave a paper describing the experiment. We reproduce some excerpts from his paper below.

> About twenty-six feet of copper wire one twentieth of an inch in diameter were wound round a cylinder of wood as a helix, the different spires of which were prevented from touching by a thin interposed twine. This helix was covered with calico, and then a second wire applied in the same manner. In this way twelve helices were superposed, each containing an average length of wire of twenty-seven feet, and all in the same direction. The first, third, fifth, seventh, ninth, and eleventh of these helices were connected at their extremities end to end, so as to form one helix; the others were connected in a similar manner; and thus two principal helices were produced, closely interposed, having the same direction, not touching anywhere and each containing one hundred and fifty-five feet in length of wire.

> One of these helices was connected with a galvanometer, the other with a voltaic battery of ten pairs of plates four inches square with double coppers and well charged; yet not the slightest sensible deflection of the galvanometer needle could be observed.

Well, that was it—total failure. It seems that a current flowing in one coil did not produce any effect in the other. Faraday tried a similar experiment using six lengths of copper wire and six of soft iron wire. Yet still no effect was observed. Then, Faraday changed the apparatus and went for more wire and more current just as any good experimenter would. Build it bigger and more powerful!

> Two hundred and three feet of copper wire in one length were coiled around a large block of wood; other two hundred and three feet of similar wire were interposed as a spiral between the turns of the first coil, and metallic contact everywhere prevented by twine. One of these helices was connected with a galvanometer, and the other with a battery of one hundred pairs of plates four inches square with double coppers, and well charged. When the contact was made, there was a sudden and very slight effect at the galvanometer, and there was also a similar slight effect when the contact with the battery was broken. But whilst the voltaic current was continuing to pass through the one helix, no galvanometrical appearances nor any effect like induction upon the other helix could be perceived, although the active power of the battery was proved to be great, by its heating the whole of its own helix, and by the brilliancy of the discharge when made through charcoal.

Faraday was persistent and made another variation of the experiment. This time he removed the galvanometer and made a small coil around a glass tube. He inserted a steel needle into the tube and found that upon removing the needle before breaking the battery contact the needle was magnetized. Then he used another unmagnetised needle but this time he made battery contact first and then introduced the needle into the tube. After breaking battery contact, the needle was removed and found to be magnetized in the opposite direction. The next variation was to place an unmagnetised needle in the tube and keep it in place while first making and then breaking the contact. This seemed to produce two contrary effects since the needle showed little or no magnetism after it was removed from the tube. Making and breaking the contact seemed to neutralize the magnetic effects.

A variation of this experiment is reproduced below using more modern equipment. A key feature is the use of an oscilloscope which shows a graph of voltage as a function of time. Too bad Faraday did not have a device like this back in his day. Even if he had a mechanical version, there is no telling what he would have discovered. Figure 1.32 illustrates an experimental arrangement in which the electrical current can be switched on and off in a manner similar to making and breaking a battery connection. Here we use a mechanical switch connected to a battery. The battery is not shown in the setup.

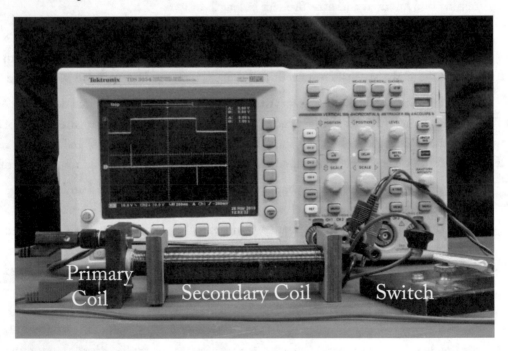

Figure 1.32: Faraday's induction experiment displayed on an oscilloscope.

The switch supplies current to the primary coil which acts as one of the helices described by Faraday. The primary coil is inserted into a secondary coil. The secondary coil represents the other helix that Faraday would have connected to his galvanometer. In place of the galvanometer, we use an oscilloscope which shows us the voltage produced in the secondary coil as a function of time.

Figure 1.33 shows the secondary coil voltage when the switch is turned on and off. The upper signal is the voltage applied to the primary coil and the lower signal is the voltage across the secondary coil. The spike in the secondary voltage appears perfectly synchronized to the switch changing its state. However, when the state of the switch remains on or off the secondary voltage drops to zero. Only when there is an abrupt change in the state of the switch do we see a voltage across the coil. We should point out that if there is a voltage across the coil that means there is a current flowing through it. When there is current flowing through the primary coil it produces a magnetic field which intersects the secondary coil. Thus, we only see a voltage across the secondary coil when the magnetic field of the primary coil changes abruptly. When the magnetic field produced by the primary coil is constant (switch on) or zero (switch off), the secondary coil voltage is zero. We can also think of this in terms of the slope of the line. Where the slope of the primary coil voltage is greatest, the signal on the secondary is greatest. When the slope of the primary is zero, the secondary voltage is also zero. This slope or derivative relationship is illustrated a little more clearly in Figure 1.34.

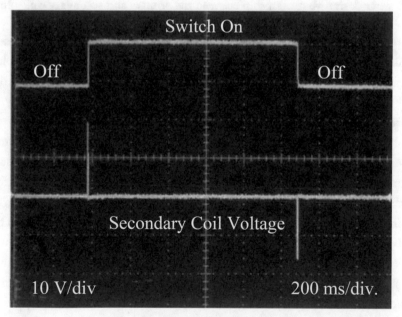

Figure 1.33: Secondary coil voltage and switch position as a function of time.

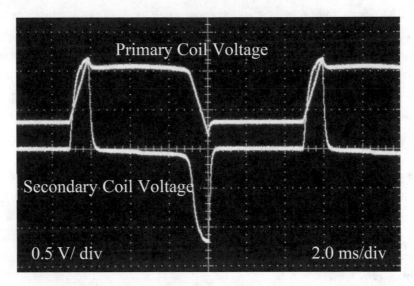

Figure 1.34: Slope or derivative relationship between the primary and secondary coil voltage.

When the primary coil voltage changes quickly the secondary coil voltage has the greatest magnitude. Likewise, when the primary voltage is nearly constant, the secondary voltage is nearly zero. This suggests that they are related by some type of rate of change or derivative nature, as illustrated in Figure 1.34. Of course, changes in the primary voltage are mirrored by the magnetic field of the primary coil. We will return to this relationship when we discuss the mathematical relationship between the magnetic field and the voltage produced. Readers with some electronics background will realize that what we have created in this experiment is the forerunner of the electrical transformer. Every time you plug an AC appliance into a wall socket, you can thank Faraday and think of this experiment. AC stands for alternating current and that is the form our electrical mains use to deliver power. The voltage on the power mains (or power lines) varies in time but with a sinusoidal shape. The exact frequency and voltage depends on where you are located. In North America, the typical single-phase voltage is 120 V at a frequency of 60 Hz. In Europe, most of Africa, and Asia, the voltage is about 230 V with a frequency of 50Hz. The fact that the secondary coil can produce a voltage driven by changes in the primary coil allows us to step up or step down the voltage used in our devices. This phenomena is also the foundation of recording information on magnetic media such as tapes and early disks. Time-varying current through a write head creates a time varying magnetic field. This field is stored in a magnetic medium such as an audio tape. To play back or read the information, one moves a magnetic medium past a read head, which consists of a coil of wire. The changing magnetic field in the vicinity of the coil induces a voltage that follows the magnetic signal that was stored in the medium.

Induction by Relative Motion

In this same paper, Faraday reports another interesting experiment in which he demonstrates that a current can be induced by relative motion. Here the current from the battery flows continuously through a wire and there is no making or breaking of electrical contact. The experiment is described below.

> Several feet of copper wire were stretched in wide zigzag forms, representing the letter W, on one surface of a broad board; a second wire was stretched in precisely similar forms on a second board, so that when brought near the first, the wires should everywhere touch, except that a sheet of thick paper was interposed. One of these wires was connected with the galvanometer, and the other with a voltaic battery. The first wire was then moved towards the second, and as it approached, the needle was deflected. Being then removed, the needle was deflected in the opposite direction. By first making the wires approach and then recede, simultaneously with the vibrations of the needle, the latter soon became very extensive; but when the wires ceased to move from or towards each other; the galvanometer needle soon came to its usual positon.

> As the wires approached, the induced current was in the contrary direction to the inducing current. As the wires receded, the induced current was in the same direction as the inducing current. When the wires remained stationary, there was no induced current.

Faraday's second experiment with the wire in the form of a "W" shows us that currents can be induced by relative motion. It does not matter which coil is moving only that the magnetic field passing through the secondary coil is changing with time. In Faraday's experiment, each wire was held fixed to a board and then one board was moved. Notice he found that there was no induced current when there was no relative motion. When the wires approached or receded the direction of the current flow changed accordingly. The wire connected to the voltaic battery produced a magnetic field. When the magnetic field at the second wire was increased in time by bringing it closer to the source of the field, an induced current flowed. When the wires were moved apart, the magnetic field at the second wire decreased in time, and once again a current was produced. This time the current flowed in the opposite direction. If Faraday had use a permanent magnet instead of a wire connected to a battery, the experiment would have produced the same result. The only thing he needed was a source of magnetic field that would cut across the wire connected to the galvanometer. Once again Faraday laid the foundation for the modern world. This experiment shows the basic principle behind the electric generator. In 1831, Faraday did create a disk-type generator in which a metal disk rotated between the poles of a horseshoe magnet. A small DC current was produced when the disk rotated. Notice it was a DC current. That is a current whose polarity does not change in time, unlike the AC current we have delivered over the power mains. Since the polarity does not change, this device is sometimes called a homopolar generator. Faraday did find that if he reversed the direction of rotation the direction of the current also reversed. Figure 1.35 shows an experimental arrangement similar to that used by Faraday. The disk is made of copper and placed between the poles of a strong permanent magnet. Electrical contact is made through the bearing and at the edge of the disk. A piece of soft solder slides along the edge of the disk to form a second contact. The disk was rotated slowly by hand and as shown in the figure, a very small current of about 1.5 μA was produced.

Figure 1.35: Reconstruction of Faraday's disk generator.

When the direction of rotation was reversed the current produced also changed sign to -1.6 μA. A few micro amps does not sound like much, but this is the first demonstration of transforming kinetic energy into a constant source of electric power. Faraday also found that he could convert electric power into kinetic energy. By doing so he invented a form of brushless motor. When you think about it a motor and a generator are just the converse of each other.

It is also possible to make a generator by rotating a coil of wire in a magnetic field. In that case you can make an AC generator. This is the type of generator that is more commonly used today at power plants. Such a generator is shown in Figure 1.36. The oscilloscope screen shows the waveform produced by the generator. What we see is not the clean sinusoidal shape that we normally have on the power mains, but it is AC power generated by a magnetic field.

It is not a stretch to say that the results of Faraday's experiments laid the groundwork for much of our modern technology. These experiments represent the beginning of the modern electrical age and opened the floodgates of technology and invention. Next time you experience a power failure take some time to think about how that electric power is produced and delivered to your community.

Figure 1.36: A small hand-powered generator. A coil rotates in the presence of a magnetic field.

1.6.4   FRANKLIN, FARADAY, AND EXPERIMENTAL SCIENCE

We have included this section on Faraday and Franklin not just to discuss magnetic fields but also to illustrate some points about how science is done. You will notice that neither Franklin nor Faraday expressed their ideas in the form of mathematical equations. Sometimes people criticize early Chinese science as not having a mathematical formulation as we have grown accustomed to in the modern day. The lack of a formal mathematical theory does not make the work of Faraday or other experimenters non-scientific. Faraday is considered by many as one of the greatest experimental physicists that ever lived. He performed many experiments that were slight variations in an attempted to understand the relationships among several variable. Faraday's work was methodical and he kept excellent notes. What he lack in mathematical training he made up for in detailed experimentation. It was up to physicists-mathematicians such as André-Marie Ampère and James Clerk Maxwell to provide the mathematical frame work of electromagnetism. There is a sort of symbiotic relationship that exists between theory and experiment. In discussing his mathematical formulation of the law of induction, Maxwell (1855) gives credit to Faraday and describes the relationship between experiment and mathematical theory as follows:

> Faraday, however, has not contented himself with simply stating the numerical results of his experiments and leaving the law to be discovered by calculation. Where he has perceived a law he has at once stated it, in terms as unambiguous as those of pure mathematics; and if the mathematician, receiving this as a physical truth, deduces from it other laws capable of being tested by experiment, he has merely assisted the physicist in arranging his own ideas, which is confessedly a necessary step in scientific

induction. In the following investigation, therefore, the laws established by Faraday will be assumed as true, and it will be shewn that by following out his speculations other and more general laws can be deduced from them. If it should then appear that these laws, originally devised to include one set of phenomena, may be generalized so as to extend to phenomena of a different class, these mathematical connexions may suggest to physicists the means of establishing physical connexions; and thus mere speculation may be turned to account in experimental science.

1.6.5   FARADAY'S LAW IN MATHEMATICAL FORM

Faraday's experiments can be explained by considering a quantity Φ_B called the magnetic flux. The magnetic flux has to do with how the magnetic field pierces through a surface. Figure 1.37 shows a magnetic field line penetrating a small patch of area ΔA and a vector that is normal (perpendicular) to the surface at that point. The angle between the normal vector and the magnetic field line is given by θ. The flux through the small patch of surface is

$$\Phi_B = B\,\Delta A \cos(\theta).$$

For a three-dimensional surface one would add up all the small patches of flux via integration, but since this is a non-calculus-based class we will deal with simple geometries such a coil of wire.

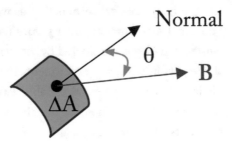

Figure 1.37: Geometry for calculating the magnetic flux through and area element ΔA.

If the magnetic field is allowed to pass through a single loop of wire and the loop is oriented perpendicular to the magnetic field, the flux through the loop is simply

$$\Phi_B = BA.$$

The unit of magnetic flux is called the weber (Wb) named after the German physicist Wilhelm Eduard Weber. Weber was an amazing physicist who in 1831 became a physics professor at the University of Göttingen at only 27 years of age. He invented an electromagnetic telegraph and was very active in the fields of acoustics and geomagnetism. The weber is given as:

$$1 \text{ Wb} = 1 \; Tm^2.$$

Faraday discovered that the electric current generated in a conducting loop is proportional to the rate of change of the flux through the loop. The electric current that flows in a circuit is driven by the electromotive force (emf) denoted by $\mathcal{E}$. Although emf has force in the name it is not actually a force. The name is a holdover from history and it is measured in units of volts. The emf actually represents a potential difference like the voltage across the terminals of a battery. We can think of emf as being related to the work done on an electric charge moving around a conducting loop. For this reason people speak of a source of emf, such as a battery, as a charge pump. Faraday's law of induction relates the emf $\mathcal{E}$ to the rate of change of the magnetic flux with time and is given by the equation

$$\mathcal{E} = -\frac{\Delta \Phi_B}{\Delta t}.$$

The minus sign in this equation indicates the direction of the emf and is related to a concept known as Lenz's law. Lenz's law tells us that the direction of the induced current is such that the magnetic field it produces opposes the change in the magnetic flux that induced the current. We can apply Faraday's law to a coil of N turns assuming that the same magnetic flux passes through all the turns equally, then the total emf induced in the coil is

$$\mathcal{E} = -N\frac{\Delta \Phi_B}{\Delta t}.$$

The magnetic flux through a coil can change several different ways. We see that the equation for Φ_B contains terms involving the strength of the field (B), the area (A) of the coil immersed in the magnetic field and the orientation (θ) between the magnetic field and the area of the coil. Changing any or all of these parameters will produce an emf and in turn a current through the coil.

Now we have a mathematical model that includes all aspects of Faraday's experiments. As we have seen demonstrated in Figures 1.33 and 1.34, the voltage across the secondary coil reacts very sensitively to changes in the magnetic field produced by the primary coil. Switching the current on and off abruptly increases the rate of change of magnetic flux ($\frac{\Delta \Phi_B}{\Delta t}$) compared to slowly varying the current through the primary coil. Keep in mind that the magnetic field of the primary coil does not change immediately in response to changes in the current. We show the secondary voltage and the rate of change of the flux through the secondary coil below in Figure 1.38.

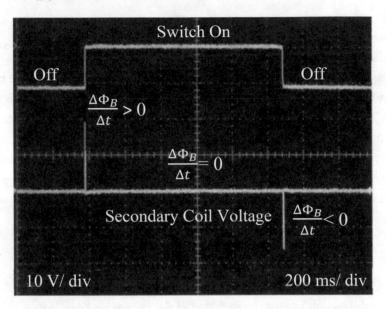

Figure 1.38: Rate of change of magnetic flux through secondary coil.

Notice that when the flux is increasing with time ($\frac{\Delta \Phi_B}{\Delta t}$ >0) the secondary voltage is positive, and when it is decreasing with time ($\frac{\Delta \Phi_B}{\Delta t}$ <0) the secondary voltage is negative. When the current through the primary is continuous ($\frac{\Delta \Phi_B}{\Delta t}$ =0), the secondary voltage drops to zero. This is completely consistent with the fact that Faraday only saw a response in the Galvanometer when he connected or broke the connection to the battery. As he described it:

> When the contact was made, there was a sudden and very slight effect at the galvanometer, and there was also a similar slight effect when the contact with the battery was broken. But whilst the voltaic current was continuing to pass through the one helix, no galvanometrical appearances nor any effect like induction upon the other helix could be perceived, although the active power of the battery was proved to be great, by its heating the whole of its own helix, and by the brilliancy of the discharge when made through charcoal.

Keep in mind that the actual sign of the voltage produced depends on the polarity of the leads connected to the oscilloscope. If we reverse the leads we would reverse the sign of the voltage.

What about Faraday's second experiment? In that case, he moved one wire with respect to the other. Again, we reproduce the key text below.

> As the wires approached, the induced current was in the contrary direction to the inducing current. As the wires receded, the induced current was in the same direction

as the inducing current. When the wires remained stationary, there was no induced current.

Now he was changing the flux through the moving wire by bringing it closer or further from the source of the magnetic field, thus varying the strength of the field cutting across the wire.

Remember the AC generator shown in Figure 1.36? How would that work? In that case, the coil of wire is rotated in the magnetic field produced by a set of permanent magnets. As the coil rotates, the flux changes for two reasons. First, the strength of the field between the magnets is not uniform. Second, as the coil rotates in the field, the angle between the coil and the magnetic field changes with time. Thus, the flux varies with time and a voltage is produced.

1.6.6   LENZ'S LAW

We will now conclude this section on induction by examining Lenz's law and how it predicts the direction of current flow through a coil. Lenz's law tells us that the direction of the induced current acts in a way to oppose the change in flux. This means that if the magnetic field though a coil is decreasing in strength, then the current will flow in such a way as to produce a magnetic field that attempts to offset that decrease. Remember that current flowing through a coil can generate its own magnetic field with the direction given by the right-hand rule. If the north pole of a magnet is moving towards a wire loop, then the flux through the loop is increasing in time. This means that $\dfrac{\Delta \Phi_B}{\Delta t} > 0$ and a current must flow in the coil to counteract the increasing magnetic flux. This situation is illustrated in Figure 1.39, which shows the north pole of a magnet approaching the loop and the loop reacting by generating its own magnetic field. The field lines are not drawn exactly to scale but only schematically to illustrate the concept. The direction of the current flow through the loop is given by a right-hand rule. If you wrap your fingers around a loop of wire in the direction of the current flow, then the thumb of your right hand will point in the direction of the magnetic field produced by the loop.

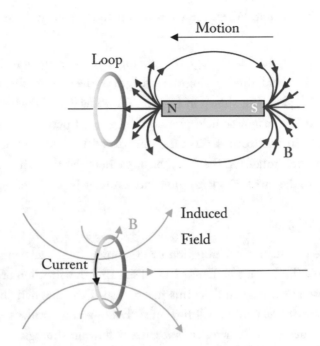

Figure 1.39: A moving magnet induces a current in a loop of wire. The induced magnetic field opposes the change in magnetic flux. The darker shaded part of the loop is further away from the observer in this two-dimensional representation.

If the magnet is moving away from the loop, then the magnetic flux is decreasing with time and $\frac{\Delta \Phi_B}{\Delta t} < 0$. The induced current in the loop will flow in such a way as to try and offset the decreasing flux by producing an induced magnetic field that adds to the field of the moving magnet. This causes the current in the loop to flow in the opposite direction. Now that we have Faraday's with additional information form Lenz's law we can understand the results Faraday obtained.

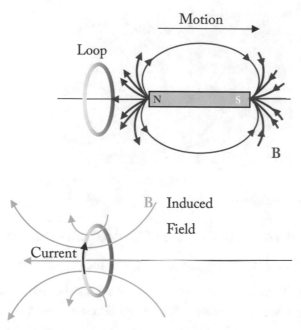

Figure 1.40: Magnet is now moving away from the loop and the induced current is reversed.

1.6.7 CALCULATING THE EMF

It will be helpful to look at a numerical example. Suppose we have a circular coil of wire whose radius is 5.00 cm and it is in a region of magnetic field that is decreasing at a rate of 5.00×10^{-2} *T/s*. The coil has five turns of wire and is oriented perpendicular to the magnetic field, as shown in Figure 1.41. The field is uniform over the area of the coil and directed into the page. A voltmeter measures the emf produced by the coil.

We can calculate the emf produced by the changing magnetic field. Since the magnetic field is oriented perpendicular to the coil, the magnetic flux would is given by $\Phi_B = BA$. The area is not changing so from Faraday's law we have

$$\mathcal{E} = -N\frac{\Delta\Phi_B}{\Delta t} = -NA\frac{\Delta B}{\Delta t} \; .$$

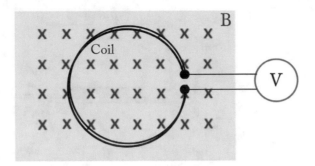

Figure 1.41: A coil of wire immersed in a time varying magnetic field.

The area of the loop is $A = \pi \times r^2 = \pi \times (5.00 \times 10^{-2} \text{m})^2 = 7.85 \times 10^{-3} \ m^2$ and the rate of change of the magnetic field is $\dfrac{\Delta B}{\Delta t} = -5.00 \times 10^{-2} \ Ts^{-1}$. The negative sign indicates that the magnetic field is decreasing. There are 5 turns which is an exact number so it can have an infinite number of significant figures, we will represent it at 3 significant figures. Substituting the values into Faraday's law yields

$$\mathcal{E} = -NA\frac{\Delta B}{\Delta t} = -5.00 \times 7.85 \times 10^{-3} \ m^2 \ x\left(-5.00 \times 10^{-2} \ Ts^{-1}\right) = 1.96 \times 10^{-3} \ Tm^2 s^{-1}.$$

The units can be simplified further to yield

$$\mathcal{E} = 1.96 \ mV.$$

To put this in perspective a AA battery is about 1.5V, so this is a very small voltage. Once you solve a problem you should always check the units. This problem has a very long string of unit conversions. The magnetic field strength expressed in Tesla can be reduced further to

$$1 \ T = \frac{1N}{A \, m}.$$

Current is expressed in units of Amps (A) and can be reduced to the rate at which charge flows. Charge is measured in Coulombs (C) so current in Amps is given by

$$1 \ A = \frac{1C}{s}.$$

Voltage (V) is related to the work done in moving charge and is expressed in Joules (J) per Coulomb. Work is measured in Joules and is related to force in Newtons (N) multiplied by a distance in meters (m). That gives us the following conversions:

$$1 \ N \ m = 1 \ J, \text{ and } 1V = 1\frac{J}{C}.$$

This is a bewildering set of units which can be expresses as follows:

$$\frac{T\,m^2}{s} = \frac{N\,m^2}{A\,m\,s} = \frac{J}{A\,s} = \frac{J}{\dfrac{C\,s}{s}} = \frac{J}{C} = V\,.$$

Notice that if the magnetic field was increasing the emf would change sign just as we expect from Faraday's experiments.

1.7   ORIGIN OF THE MAGNETIC FIELD

The origin of the magnetic field is a very complex topic and the full explanation requires principles from quantum mechanics. We will examine a simplified way of thinking about magnetic fields that does not capture all of the quantum mechanical details. Ørsted's experiment shows us that an electrical current produces a magnetic field. Shortly after that discovery, the French polymath, André-Marie Ampère, began work on a mathematical theory to relate electricity and magnetism. Ampère was not only an excellent theorist but he was also a highly skilled experimentalist. His name may sound familiar, since the unit of electrical current the Ampere is named after him. He is also credited with giving us the name solenoid to describe the coil we discussed previously (see Figure 1.29).

One of Ampère's greatest experimental achievements was the discovery that two parallel conductors carrying currents can be attracted or repelled depending on the direction of the current flowing through them. These results led Ampère to theorize that all magnetic interactions were due to the currents flowing in conducting materials. This, he thought, was true on all size scales. Thus, he believed that there were electric currents flowing through the Earth. He also thought that within magnetic materials such as 慈石 císhí, there were macroscopic electric currents. He expressed this idea in several different ways and we reproduce one such statement below (Ampère, 1820).

> Now, if electric currents are the cause of the directive action of the Earth, then electric currents could also cause the action of one magnet on another magnet. It therefore follows that a magnet could be regarded as an assembly of electric currents in planes perpendicular to its axis, their direction being such that the austral pole of the magnet, pointing North, is to the right of these currents since it is always to the left of a current placed outside the magnet, and which faces it in a parallel direction, or rather that these currents establish themselves first in the magnet along the shortest closed curves, whether from left to right, or from right to left, and the line perpendicular to the planes of these currents then becomes the axis of the magnet and its extremities make the two poles. Thus, at each pole, the electric currents of which the magnet is composed are directed along closed concentric curves.

Initially, Ampère's current flow model invoked macroscopic currents flowing around the inside of the magnet. As a result of his discussions with his friend Augustin-Jean Fresnel, of optics fame, Ampère modified his theory. Fresnel pointed out that, if one made a magnet out of a hollow steel cylinder, Ampère's theory predicted that currents would flow around the axis of the cylinder. If we cut the cylinder along the long axis, this would interrupt the current flow and the magnetism would be destroyed. This, however, is not what we observe experimentally. Fresnel suggested that the currents must be microscopic and flowing around each particle of iron that makes up the magnet. Fresnel also pointed out several other experimental facts that ruled out the idea of macroscopic currents but allowed for the possibility of microscopic currents.

Interestingly enough, Ørsted and many notable scientists opposed Ampère theories and attempted to create their own theoretical explanations. We mention this just to reiterate the idea that fully formed theories do not just fall out of the sky.

Ampère's theory that there were currents flowing on the microscopic scale was insightful. At the time Ampère was conducting his experiments the electron had not yet been discovered. One way to explain to origin of magnetism in materials is through a model in which the circulation of the electron about the nucleus is treated as a current. Not only does the electron orbit around the nucleus, it also has an inherent angular momentum. This leads to a quantity known as spin. The concept of electron spin was introduced in 1925 by Dutch physicists Goudsmit and Uhlenbeck. Interested readers can find more information about electron spin as described by Samuel Goudsmit at https://www.lorentz.leidenuniv.nl/history/spin/goudsmit.html.

The spin and orbital properties of the electron are vector quantities known as the spin magnetic dipole moment and orbital magnetic dipole moment. These quantities add together vectorially much the same way force vectors can be added together to give a resultant vector. In a piece of magnetic material the vectors components corresponding to all the electrons in that material add together. In some cases, the electrons in the atom may be paired. These paired electrons spin and orbit in opposite directions and their magnetic moments cancel out. For a material made up of completely paired electrons, the magnetic fields of all the atoms roughly cancel out.

If you place a material in an external magnetic field, the magnetic moments of the atoms will respond somewhat like compass needles and orient themselves. The external field then induces a net dipole moment in the material. We classify different materials by how this induced magnetic field behaves and the overall effect if has on the material. Some materials are diamagnetic and they form magnetic dipoles that are oriented opposite to the external magnetic field. Since they respond by pointing in the opposite direction, a diamagnetic material will repel both ends of a simple bar magnet. Perhaps the most common diamagnetic material is just plain old water. The effect is slight and a large magnetic field is often necessary to observe the repulsion. Living creatures, containing water, have been levitated in intense magnetic fields. A good example can be found at https://www.youtube.com/watch?v=KlJsVqc0ywM. In this video, a frog and several other objects are placed inside a solenoid and subjected to an intense magnetic field. Bismuth, gold, mercury, and zinc are also diamagnetic. You may have seen images of magnets levitating over superconductors. In addition to exhibiting no electrical resistance superconductors are diamagnetic.

Aluminum has the electron configuration [Ne] $3s^2 3p^1$, so, there is one unpaired electron. This unpaired electron enables aluminum to become magnetized. But, you are thinking that aluminum is not magnetic like iron. That is true. Aluminum is an example of a paramagnetic material. Transition elements, rare earths, and actinides all exhibit a quality known as paramagnetism. The atoms that make up a paramagnetic material are all randomly oriented and their magnetic moments are all pointing in random directions. If you apply an external magnetic field to a paramagnetic material, the dipole moments partially line up with the applied field. This causes the material to be attracted to a strong permanent magnet. If you remove the external magnetic field, the partial alignment disappears. Apart from metals, liquid oxygen is also paramagnetic and can be suspended between the poles of a strong magnet.

Suppose you apply an external magnetic field to a piece of nickel or iron? The magnetic dipole moments in these materials will align more completely with the applied magnetic field. If you remove the applied magnetic field, the material will retain some of its alignment and remain permanently magnetized. Ferromagnetic materials often contain magnet domains. In a magnetic domain, small regions of the material have dipole moments pointing in similar directions. Adjacent domains have magnetic dipole moments pointing in different directions. When an external magnetic field is applied, the magnetic dipole moments try to line up with the applied field. Magnetic domains that were initially oriented with the applied field tend to grow in size. Other domains attempt to align themselves with the applied magnetic field. The overall alignment produces a strong magnetic response in materials such as iron. This is shown schematically in Figure 1.42.

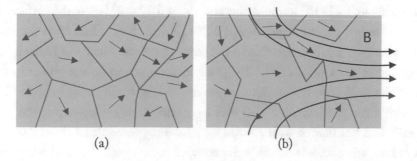

(a)   (b)

Figure 1.42: A schematic representation of magnetic domain in a ferromagnetic material. The magnetic moment is shown by the dark arrow. (a) No external applied magnetic field and (b) in an applied magnetic field.

An excellent animation of magnetic domains growing in the presence of a magnetic field can be found at https://en.wikipedia.org/wiki/Magnetic_domain#/media/File:Moving_magnetic_domains_by_Zureks.gif.

When we say ferromagnetic materials are permanently magnetized, that does not quite mean forever. If you strike a magnetized bar of iron with a hammer, you can randomize the magnetic orientation and demagnetize the material. Another way to randomize the magnetic dipole moments is by thermal energy. If you raise the temperature of a ferromagnetic material above the Curie temperature for that particular material, the degree of alignment decreases. Usually, the material then becomes paramagnetic with a weak alignment. For nickel, this occurs at about 354°C, while the Currie temperature for iron is about 770°C. The Currie that discovered this effect is Pierre, husband of the famous Marie.

1.8   THE MAGNETOSPHERE OF THE EARTH

In this last section we want to return to 沈括 Shěn Kuò's initial observations of the Earth's magnetic field. We have shown that a magnetic field is produced when a current flows. Does the Earth contain a huge battery? This is a very good question and geophysicist have been asking the same sort of question for several hundred years. In 1838, the great German mathematician and physicist Carl Friedrich Gauss created a mathematical model, using spherical harmonics, to describe the global magnetic field. In his model, Gauss separated out internal and external components of the Earth's magnetic field. The dominant component of Earth magnetic field is called the core or main field and is thought to be generated by a hydrodynamic dynamo in the fluid core of the Earth. All of the details of the source of Earth's magnetic field are not fully understood. Seismic data suggests that the Earth has a solid inner core which is surrounded by a molten core. The inner core is thought to be about 1.22×10^3 km in radius and about 6.37×10^3 below the surface. Surrounding the solid inner core is a molten region about 2.30×10^3 km thick. Heat from the inner core and compositional gradients create convection in the outer core fluid. Since the Earth is rotating there is also mixing due to the Coriolis force. The geodynamo needs to be self-sustaining. Simulations indicate that if the field were not continuously generated, it would decay in about 2.0×10^4 years. Another feature of the geodynamo is that over long periods of geologic time the polarity of the Earth magnetic field has flipped with a mean time between reversals of about 2.0×10^5 years. The exact details of how this occurs is still an active research are in geophysics and many researchers are engaged in creating numerical simulations to understand the process.

Earlier in this chapter we found that the World Magnetic Model predicted a magnetic field strength of 4.998×10^4 nT at 镇江市 Zhènjiāngshì. At the Earth's surface the main field is in the range of about 2×10^4 nT to 7×10^4 nT. A second contribution to the magnetic field is that of the lithosphere or crustal field. This contribution comes from magnetized rocks such as the 慈石 císhí or lodestone we encountered in the beginning of the chapter. The crustal field varies considerably over the surface of the Earth with variations on the order of a few hundred *nT*.

There are also external components that contribute to the Earth's magnetic field. These include the ionosphere and the magnetosphere with the Sun. The ionosphere consists of ionized atoms molecules extending from a height of about 50–1,000 km above the Earth. Surrounding our planet is a region of space that interacts with charge particles from the Sun. This region is known as the magnetosphere. The magnetosphere extends out into space and it shape is distorted by the solar wind, as shown in Figure 1.42.

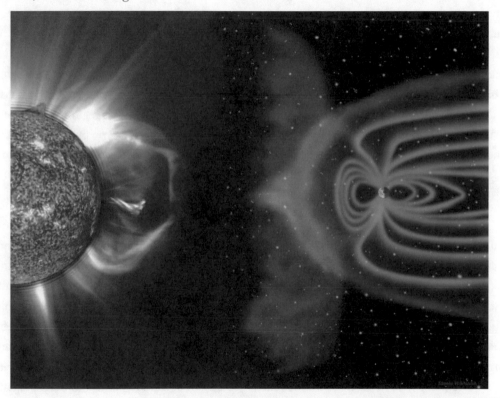

Figure 1.43: Artist's Conception of the magnetosphere of the Earth interacting with high energy particles from the Sun, via images-assets.nasa.gov/image/0201490/0201490~orig.jpg. Not to scale.

Yes, there is wind in space. The solar wind consists of charged particles such as electrons protons and alpha particles emitted from the Sun with velocities of several hundred km/s. Trace amounts of heavy ions and atomic nuclei are also found in the solar wind. Pressure from the solar wind compresses the magnetosphere on the side towards the Sun. On the Sun side the magnetosphere extends about 10 Earth radii and on the opposite side it is elongated out to more than 100 Earth radii. The external contributions to the magnetic field near the surface of the Earth account for about 20 *nT* in the mid latitudes and can spike up several orders of magnitude during a magnetic storm. Activity in the ionosphere and magnetosphere are directly driven by events on the Sun. For up-to-date solar wind measurements and general information about space weather, interested readers should consult the Space Weather website at found at https://www.spaceweather.com/.

There is even a small component of Earth's magnetic field due to the ocean currents. This ocean tidal magnetic field was clearly demonstrated by satellite mapping data at the beginning of the 21st century (Tyler et al., 2003). The ocean tidal dynamo only accounts for a few *nT* but its origin is reminiscent of Faraday's law on a global scale. The current model explains the field as being generated by the movement of highly conductive seawater through Earth's magnetic field. This produces induced electric currents, which then generate secondary magnetic fields.

Stars and planets can generate magnetic fields. Jupiter, Saturn, Uranus, and Neptune all have magnetic fields which are greater than the magnetic field of the Earth. Jupiter's magnetic field is very intense at about 1.95×10^4 times that of the Earth. Venus does not have a magnetic field and Mercury has a very weak field about 6×10^{-3} times that of the Earth. Mars has regions of magnetized material in the crust but no well-formed global magnetic field. There is a type of Neutron star that is believed to produce magnetic fields of $10^9 - 10^{11}$ *T* about 19 order of magnitude greater than the Earth's magnetic field.

1.9   WHEN DOES IT BECOME SCIENCE?

When does it become science? That is the question that often accompanies our study of Ancient Chinese Science and perhaps all of the history of science. Was it science when Faraday, Franklin, Ørsted, and Henry were conducting their experiments? Modern science demands a type of cultural liberation where you lift an idea out of 五行 wǔxíng or flow of the 气 qì and express your explanation in a more universal language. Some would tell us that an idea becomes science when you can express it mathematically. Faraday was not a mathematician and it was up to James Clerk Maxwell to formulate the mathematical theories that explained Faraday's work. In his paper, *On Faraday's Lines of Force* (1855), he applied the mathematical formalism of fluid dynamics to derive equations relating to electromagnetism. Yet, that was more than 30 years after the experimental work. Did induction become science only when it was expressed mathematically? Even after the mathematical descriptions of electromagnetism were demonstrated, we still did not understand what was carrying the charge in the induced currents. It was not until 1897 that British Physicist Joseph John Thomson discovered that cathode rays were composed of the negatively charged particles that we now call electrons. The point is that we knew currents produced fields but we did not understand the nature of the particles producing the currents. This is the progression of knowledge that we call science.

We have mentioned the solar wind. When did it become science? The *Book of Jin* 晋书 Jìn Shū is an official history of the Jin Dynasty or 晋朝 Jìn Cháo. In the 晋书 Jìn Shū , which covers the years 265–420, we find a statement directly related to the solar wind. This ancient text tells us that the tail of a comet always points in the direction of light radiating from the sun. It does not matter if the comet is in the northern part of the sky or the southern part of the sky, the tail always points away from the Sun. The ancient Chinese were great at recording the position and shape of comets and this observation was reproducible. Some comets do not develop a tail and only have a nucleus. If they do have a tail, it always points away from the Sun no matter the direction of the comet in its orbit. Why do they point this way? An ancient Chinese astronomer would say this is a direct consequence of the 气 qì emitted from the Sun deflecting the tail of the comet. In 1532, Petrus Apianus, a German mathematician and astronomer, realized that comet tails always lie on a line that passes through the Sun. In 1910, British Astrophysicist Arthur Eddington postulated the existence of electrons ejected from the Sun. He did not use the term solar wind but suggested that charged particles from the Sun interact with comets. We now know that comets often exhibit two tails composed of different types of matter. One tail is primarily made of dust and small particles that are liberated from the coma or head of the comet as is heats up in the vicinity of the Sun. These dust particles reflect sunlight and become visible. The second tail is made up of ions. Gasses ejected from the coma are ionized by short wavelength (UV) light. Electrons are ejected from the gas molecules thus forming ions. These ions interact with the magnetic field of the Sun and experience a force which deflects the ion tail. The dust and ion tails deflect by different amounts and this results in a splitting of the tail into two components, as shown in Figure 1.44. When did our understanding of this phenomena become scientific? Was it in the 晋朝 Jìn Cháo, the 16th century, or in the 20th century? The science of heliophysics has advance greatly in the past century, yet there are still unanswered questions regarding the Sun and its magnetosphere, as evidenced by the collage of NASA heliophysics missions shown in Figure 1.45.

Figure 1.44: Photograph of comet Lovejoy near the Earth's horizon taken from a unique perspective on board the International space Station by Dan Burbank, Expedition 30 commander, December 22, 2011, via http://spaceflight.nasa.gov/gallery/images/station/crew-30/html/iss030e015472.html.

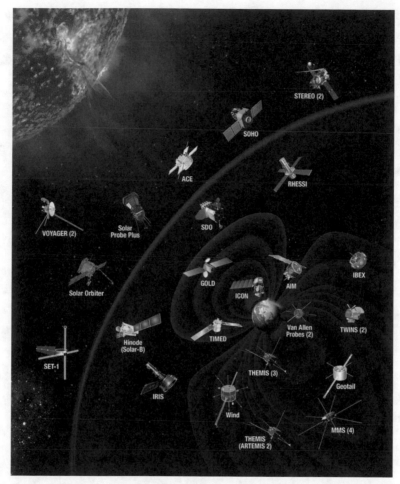

Figure 1.45: **Collage of NASA heliophysics missions,** via https://www.nasa.gov/sites/default/files/styles/full_width_feature/public/thumbnails/image/hso_image1.jpg.

REFERENCES

Ampère, A.-M. (1820). Suite du Mémoire sur l'Action mutuelle entre deux courans électriques, entre un courant électrique et un aimant ou le globe terrestre, et entre deux aimants. *Annales de Chimie et de Physique*, 15, 170–218. 47

Bacon, F. (1960). *The New Organon and Related Writings* (Library of Liberal Arts, no. 97). (F. H. Anderson, Ed.) Prentice Hall. 1

Franklin, B. (1769). Experiments and Observations on Electricity, Made at Philadelphia in America by Benjamin Franklin. London: David Henry. 18

Franklin, B. (1970). *The Writing of Benjamin Franklin* (Vol. VI). (A. H. Smyth, Ed.) New York: Haskell House. 19

Good, R. P. (1987). *A Brief History of Geomagnetism and A Catalog of the Collections of the National Museum of American History. 2.* 3

Maxwell, J. C. (1855). On Faraday's Lines of Force. Transactions of the Cambridge Philosophical Society, X, 155-229. DOI: 10.1017/CBO9780511698095.011. 39, 52

Needham, J., Wang, L., and Robinson, K. (1962). *Science and Civilisation in China, Part 1 Physics* (Vol. 4). Cambridge University Press. 2

Ørsted, H. C. (1820). Experiments on the Effect of a Current of Electricity on the Magnetic Needle. *Annals of Philosophy*, XVI, 273–276. 28

Sturgeon, W., Christie, S., Gregory, O., and Barlow, P. (1824). No. III. IMPROVED ELECTRO-MAGNETIC APPARATUS. *Transactions of the Society, Instituted at London, for the Encouragement of Arts, Manufactures, and Commerce*, 43, 37–52. Retrieved January 13, 2020, from www.jstor.org/stable/41325678. 30

Tyler, R. H., Maus, S., and Luhr, H. (2003). Satellite observations of magnetic fields due to ocean tidal flow. *Science*, 299(5604), 239–41. Retrieved from https://search.proquest.com/docview/213598955?accountid=12381. DOI: 10.1126/science.1078074. 52

CHAPTER 2

Spherical Mirrors
球面镜 Qiúmiànjìng

2.1 MIRRORS AND SYMBOLISM

In ancient China the mirror was more than a device for observing one's own image. Mirrors held special and mystical powers. Mirrors were used in ritual magic and the decoration on them were designed to invoked cosmic powers. In this chapter we will explore the physics of image formation as well as some the metaphysical properties attributed to mirrors. Ancient China was well known for their mastery of bronze casting and the mirrors we will discuss brought bronze casting to its highest art. One of the oldest bronze mirrors found dates back to about 2000 BC (Zhu and He, 1993). This ancient mirror was found in Guinan county (贵南县 Guìnánxiàn) which is located in the Hainan Tibetan Autonomous Prefecture (海南藏族自治州 Hǎinán Zàngzú Zìzhìzhōu) also known as མཚོ་ལྷོ་བོད་རིགས་རང་སྐྱོང་ཁུལ.

The mirror shown below in Figure 2.1 is from the author's collection. This mirror is an example of the lion and grape design which was very popular during the 唐朝 Tángcháo, or Tang Dynasty (618–906). Unfortunately, this design is also one of the most commonly counterfeited and so the authenticity of this particular mirror is questionable. Like all Chinese mirrors, the example shown in Figure 2.1 is full of symbolism. Throughout this section we will mix discussions of optics with some of the folklore surrounding mirrors. In many ancient cultures mundane objects are often imbued with metaphysical meaning to remind the user of some particular lesson. In our modern world a mirror is just a mirror. The mirror shown in Figure 2.1 is round. That seems to have the obvious reason of being held in the hand. However, to the ancient Chinese, round was associated with the heavens. Round shapes symbolized heaven and square shapes symbolized the Earth. There are square Chinese mirrors, mirrors with eight lobes, six lobes, and other unusual shapes. All of these shapes and patterns are designed to tell a story. They may invoke blessings, ward off evil, or provide connection to the heavens. Mirrors buried in tombs were thought to provide light in the dark underworld. Around the year 640, the Tang court began importing grapes from Central Asia. According to sinologists such as Schuyler Cammann, grapes and wine become symbolic of eating, drinking, riches, and plenty during times of universal peace (Cammann, 1953). We also see that there are two peafowl with some differences. Male and female birds positioned symmetrically on a mirror represent the unity of marriage. Sometimes this type of mirror would be hung over the couple's bed to ward off evil influences. The Chinese phoenix was a popular symbol on marriage mirrors since it was believed to monogamous. Peacocks or 孔雀 kǒngquè are thought to have a fiery nature and are sometimes used to revive the fires of love in a relationship.

Figure 2.1: A bronze lion and grape marriage mirror. This design was common in the 唐朝 Tángcháo, or Tang Dynasty (618–906).

Underneath the central lion on this mirror there a small hole so that a red ribbon can be attached. The ribbon allows the user to hold the mirror without getting their fingers on the polished surface. There is more to the central lion than just a support point. Center or 中 zhōng has important cosmological or religious significance. In some forms of Chinese cosmology, the North Star or the North Celestial Pole, is a symbol of the 阊阖 chānghé or the Heavenly Gate. Through this aperture, the Lord of Central Heaven, sometimes known as 黄帝 Huángdì, or the Yellow Emperor, could observe the Earth and its inhabitants. Prayers and sacrifices were believed to ascend to him through the Heavenly gate. Thus, the central position of a mirror was like some sort of conduit connecting life on the Earth to the Heavens (天 tiān). One final bit of symbolism is connected to the fact that there are seven lions. This is likely to be a reference to 日月五星 rì yuè wǔ xīng which signifies the five visible planets plus the Sun (日 rì) and Moon (月 yuè). The five visible planets are: Mercury 水星 (Shuǐxīng), Venus 金星 (Jīnxīng), Mars 火星 (Huǒxīng), Jupiter 木星

(Mùxīng), and Saturn 土星 (Tǔxīng). Each planet name has the character 星 xīng which means star. Their names also contain the characters for the 五行 wǔxíng known as the five phases or the five elements. These are: water 水 (shuǐ), metal 金 (jīn), fire 火 (huǒ), wood 木 (mù), and earth 土 (tǔ). This little 7-cm-diameter bronze disk holds entire classes on Chinese cosmology, philosophy, folklore, metallurgy, and optics. We shall now dive into the optics with a few more side trips into Chinese metaphysics.

Readers interested in Chinese bronze mirrors may enjoy the collection at the Harvard Art Museum which can be accessed on line at https://www.harvardartmuseums.org/collections. A highly ornate example Tang dynasty lion and grape mirror can be found at the following link: https://www.harvardartmuseums.org/collections/object/204116?position=0.

2.2   CONCAVE MIRRORS DESCRIBED IN THE 墨子 MÒZǏ

In *Teaching Physics through Ancient Chinese Science and Technology* we introduced one of the most important texts in the history of Chinese logic and science known at the 墨子 (Mòzǐ). Students of Chinese philosophy are familiar with this text for its discussions of ethics, defensive fortifications, and analogical argumentation. Yet it contains numerous short statements related to concepts of space, time mechanics, geometry, and optics. In this chapter we will explore some of what the 墨子 Mòzǐ teaches us about optics and spherical mirrors. For readers unfamiliar with the 墨子 Mòzǐ we should point out that much of the book is attributed to 墨子 (Mòzǐ) who is also known by the name 墨翟 (Mò Dí). We say "Attributed to" because it is not entirely clear how much of the work is original with 墨翟 Mò Dí and what parts were written by his followers. In the West he is sometimes referred to by the Latinized version of his name which is Micius. The dates of his life are not certain, but he probably lived c. 470–391 BC. This would make him a contemporary of Socrates. In Ancient China this era was known as the Waring States period (战国时代 Zhànguó Shídài). Ancient Chinese uses very compact language and this make translation very difficult. There are several English translations of the 墨子 Mòzǐ and, not surprisingly, they all read very differently. The translation problem is compounded by the fact that the text uses many characters whose meaning is lost or uncertain. Thus, copyists may have introduced new or erroneous meanings. This has led to numerous emendations. For all of its importance in the history of Chinese Science it seems to be one of the most difficult works to accurately translate. Needham (Needham and Wang, 1956) and his colleagues have translated some sections and they can be found in *Science and Civilisation in China*. Two other important translations are by A.C. Graham (1978) and Ian Johnston (2010).

In Book 10 (卷十 Juàn Shí) of the 墨子 Mòzǐ we find a section which is sometimes translated as Cannons (经 Jīng) and Explanations (经说 Jīng shuō). The cannon (经 Jīng) consists of short compact statements having to do with language, logic, and fragments of scientific thought. Each Canon is followed by an explanation, (经说 Jīng shuō) the purpose of which is to further explain or clarify the statement. There are two sections of Canons and Explanations 經上 (Jīng shàng) and 經下 (Jīng xià) or Cannon A and B as we will call them. Unless otherwise noted, we follow the Johnston translation (Johnston, 2010), which contains Cannon A1–A99 and B1–B81. The English text is given with the Cannon reference number in bold face.

2.2.1   CONCAVE MIRRORS

The 墨子 Mòzǐ includes some very interesting descriptions of concave spherical mirrors which we will explore in detail. Our objective is to compare what is described in the 墨子 Mòzǐ to our modern understanding of optics and ray tracing. In the 墨子 Mòzǐ we read the following (B23).

> In a concave mirror there are two images; one is small and inverted and one is large and upright. The explanation lies in weather the object is outside or inside the center of curvature.

> A mirror: When the object is within the center and approaches the center, that what is reflected becomes larger and the image is also larger. When it moves away from the center, then what is reflected becomes smaller and the image also smaller, yet necessarily upright. Arising at the center is the cause of being upright and extending its vertical height. When the object is outside the center and approaches the center then what is reflected becomes larger and the image also larger. When it moves away from the center, then what is reflected becomes smaller and the image is also smaller, and inverted. Converging at the center is the cause of being changed and extending its vertical height.

2.2.2   REFLECTION AND PLANE MIRRORS

Before we attempt to understand what is being described in the 墨子 Mòzǐ we need to establish a few basic facts about mirrors and reflection. Suppose you are standing in front of a large plane mirror such as in a bathroom or a fitting room. Things in front of the mirror appear to be behind the wall that the mirror is mounted on. In fact, people often use mirrors to give the illusion that a small room is larger than it actually is. As we will see, an object located in front of a plane mirror at a distance of 1.0 m looks like it is behind the mirror at a distance of 1.0 m. Similarly, an object at 3.0 m in front of the mirror appears to be 3.0 m into the wall. When we see an image formed by reflection off of a mirror, or brain interprets the reflected light rays as if they originated from a point along a straight line connecting directly to our eye. So, even though the source of the light ray is on the same side of the mirror as the observer, the brain interprets the image as originating from a point behind the mirror. This is shown below in Figure 2.3. The other basic fact we need to understand is the law of reflection. If a ray of light is incident on a flat reflecting plane, then the angle of the reflected ray is equal to the angle of the incident ray as measured with respect to a line normal to the surface. The law of reflection is illustrated in Figure 2.2.

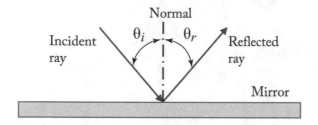

Figure 2.2: Illustrating the law of reflection in which the angle of the reflected ray (θ_r) is equal to the angle of the incident ray (θ_i).

Now we can apply this law to a tall plane mirror with an observer standing in front of it and looking at some object as shown, in Figure 2.3. Here the object is an arrow placed in front of the mirror. The light rays from the top and bottom of the object strike the mirror and reflect off such that their reflected angles are equal to their incident angles. Throughout this chapter we will indicate the incident angle as θ_i and the reflected angle as θ_r. The dark black arrow indicates the object and the lighter gray arrow indicates the image. We use an arrow so that we can determine the relative orientation of the object and image. In some cases the image may be inverted. This is shown by having the image arrow pointing down while the object arrow is pointing up. The relative lengths of the arrows indicate the relative size of the image compared to the size of the object. Some optical configurations produce an image that is shorter than the object and some produce an image large than the object. These lengths are represented numerically by h_i and h_o. For the plane mirror shown in Figure 2.3 the object arrow is pointed up and the image arrow also points up. The dashed line represents an axis that is normal or perpendicular to the plane of the mirror at the point that the incident ray strikes or is reflected from the surface of the mirror. Solid red lines represent light rays on the same side of the mirror as the object. Dashed red lines represent lines that are extrapolations of the light rays. The distance of the object from the plane of the mirror is given by d_o and the distance of the image is given by d_i. With these conventions in mind let us explore the simple plane mirror. Applying the simple law of reflection to the mirror we can construct the ray diagram shown below in Figure 2.3. We use two incident rays, one from the base of the object arrow and one from the tip. This allows us to predict the orientation of the image.

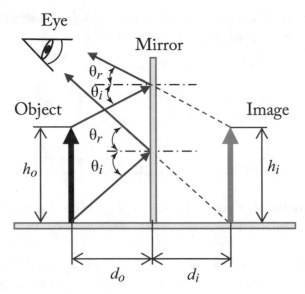

Figure 2.3: Image formed by a plane mirror.

Our brain extrapolates the reflected light rays in straight lines as if they originated from some position behind the mirror. The distances d_o and d_i are known as the object and image distances. For the plane mirror these distance are equal. The height of the object and the image are given by h_o and h_i which are also equal in this situation. This is consistent with what we observe when looking into a flat vanity mirror something like what you might find in a fitting room or a bathroom. The image seems to exist in the wall behind the mirror. That image seems to be positioned behind the mirror by a distance equal to the object distance. Objects further away from the mirror seem to recede into a space further behind the mirror. In reality, the only thing behind the mirror is the wall. This optical situation produces the appearance of depth and the image is described as being virtual.

2.2.3   SPHERICAL MIRRORS

Plane mirrors are simple to understand, but the mirror described in the 墨子 Mòzǐ has two images. One image is small and inverted and the other image is large and upright. This implies that we are dealing with a concave or 凹面镜 āomiànjìng. The character 镜 jìng is the general term for a mirror. 凹 āo looks just like what it means, concave. 面 miàn means surface. An 凹面镜 āomiànjìng is a type of 球面镜 qiúmiànjìng or spherical mirror. 球 qiú means a ball or sphere. The geometry of a concave mirror is shown in Figure 2.4. The reflecting surface is shaped like a section of a sphere with a radius of curvature R. The center of the sphere is indicated by C and the focal length is given by f. For a spherical mirror of radius R, the focal length f is given by

$$f = \frac{R}{2}.$$

This equation is only approximately true. It works well for rays that strike the mirror near the center, so that the incident and reflected angles are small. Light rays traveling parallel to the optical axis reflect off the surface and pass through the focal point, as shown in Figure 2.4. Notice that the angle of incidence θ_i is equal to the angle of reflection θ_r. The line C-N is along the radius so it is normal (perpendicular) to the surface of the mirror.

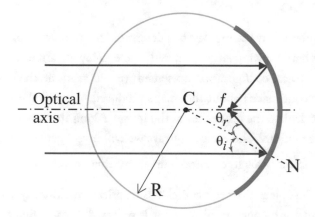

Figure 2.4: Geometry of a concave spherical mirror with radius of curvature R and focal length f.

A second type of spherical mirror is known as 凸面镜 tūmiànjìng or a convex mirror. 凸 tū is a protruding shape and in this case has the meaning convex. A convex mirror is shown in Figure 2.5.

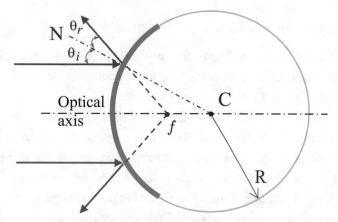

Figure 2.5: Geometry of a convex spherical mirror with radius of curvature R and focal length f. Diverging rays appear to come from a virtual source located at the focal length behind the mirror.

Light rays parallel to the optical axis strike the mirror and diverge as shown in Figure 2.5. If we extrapolate the reflected rays backward through the mirror (dashed lines), then they do converge at the focal point. The point at which they converge is behind the mirror, somewhat like a plane mirror. Thus images formed by such a mirror are virtual. We will now examine several different types of optical configurations that use spherical mirrors and see how they compare to what we have learned from the 墨子 Mòzǐ.

2.2.4   RAY TRACING

There are three basic rules of ray tracing. We can determine the properties of an image by using any two rays and noting where they intersect. Generally, we employ all three rules to be certain of our conclusions. For each optical configuration presented we will examine the behavior of each ray to draw some qualitative conclusions about the image produced. We will then employ mathematical equations to actually calculate the properties of the image. Along the way we will make some connections to the Chinese culture. The important rays we will investigate are as follows.

- **Ray 1:** From object through center of curvature. This ray reflects back on itself.

- **Ray 2:** From the object directed parallel to the optical axis. This ray is reflected through the focal point for a concave mirror or it is reflected along a line that extends away from the focal point of a convex mirror.

- **Ray 3:** From the object along a path that goes through or in the direction of a line that passes through the focal point. This ray is reflected parallel to the optical axis.

Ray Tracing Applied to a Concave Mirror

Now we will apply the rules of ray tracing to the concave mirror shown in Figure 2.6. We use the tip of an upright arrow to indicate the object and then try to determine the orientation of the image. The image is represented by the tip of a lighter gray arrow.

Ray 1 travels from the tip of the object through the center of curvature. This ray strikes the mirror and reflects back on itself. Note that the rays are extended back to the line marked "M" which represents the plane of the mirror and not arc that represents the mirror's surface. This is the usual convention in ray tracing. Now we allow Ray 2 to travel parallel to the optical axis and strike the mirror. This ray is reflected along a line that must pass through the focal point. These two rays intersect at a point which is below the optical axis and we immediately see that the image will be inverted. Ray 3 is directed along a line that passes through the focal point. Consequently, the reflected ray travels back along a path that is parallel to the optical axis. The image is formed where the rays intersect. As expected, the image is inverted and smaller than the object. Notice the rays we used were at the tip of the object and the intersection point is the tip of the image.

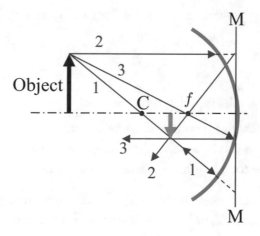

Figure 2.6: Ray diagram for a concave mirror that produces a real and inverted image.

Calculating the Image Distance for a Concave Mirror ($d_0 > f$)

Ray tracing is a powerful technique that gives us an idea about the size and shape of the image. We can also calculate the properties of the image. These calculations will give us the location of the image based on the distance between the object and the mirror. From the object and image distances we can calculate the ratio of the image height to object height. If we know the height of the original object, we can calculate the actual height of the image. We will now analyze the properties of the image produced by the concave mirror shown in Figure 2.6. Figure 2.7 shows the important mathematical parameters.

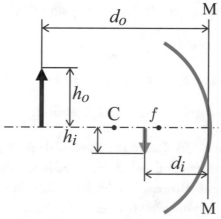

Figure 2.7: Method for calculating the image distance and height for a concave mirror. The object is placed at a distance greater than the focal length ($d_0 > f$) .

We can calculate the position of the image from the focal length and the position of the object using the equation

$$\frac{1}{f} = \frac{1}{d_o} + \frac{1}{d_i} \, .$$

There are two general rules for using this equation.

- **Rule 1:** The focal length f is positive if the focal point is on the same side of the mirror as the incident ray and it is negative if it is on the opposite side. This means that for a concave mirror the focal length is positive and for a convex mirror it is negative.

- **Rule 2:** The object and image distances (d_o, d_i) are positive if they are on the same side of the mirror as the incident light ray. If they are on the opposite side of the mirror, then they are negative. A negative image distance indicates that the image is virtual and a positive value indicates that it is a real image.

The height of the object (h_o) is related to the height of the image (h_i) by the equation

$$m = -\frac{d_i}{d_o} = \frac{h_i}{h_o} \, ,$$

where m is called the magnification.

Let's apply these equations to the mirror shown in Figure 2.7 with $f = 10.0$ cm and $d_o = 34.5$ cm.

$$\frac{1}{10.0 \, cm} = \frac{1}{34.5 \, cm} + \frac{1}{d_i} \, ,$$

which yields an image distance of $d_i = 14.1$ cm.

From the image distance we can calculate the magnification and the height of the image:

$$m = -\frac{d_i}{d_o} = \frac{14.1 \, cm}{34.5 \, cm} = -0.409 = \frac{h_i}{h_o} \, .$$

If the height of the object is $h_o = 12.0$ cm, then the height of the image is

$$m \times h_o = h_i = -4.09 \times 12.0 \, cm = -4.91 \, cm \, .$$

The negative sign indicates that the image is inverted. From the ray tracing method, we have already shown that the image is smaller than the object and inverted. Our mathematical results are perfectly in keeping with the qualitative results from ray tracing. They are also in line with what the 墨子 Mòzǐ indicates. Remember the passage instructs us that there will be two types of images, one of which is small and inverted.

In ancient Chinese culture, this particular optical configuration was thought to have great mystical powers. Mirrors were used to ward off ghosts or evil spirits. It was a common belief that if a spirit saw its own image it would be frightened or feel that it could no longer hide. A concave mirror was especially useful for this purpose. People thought that if a spirit saw its own image it would be frightened to see itself flipped upside down and diminished in size by a mirror. In fact, some Daoist adepts would wear such mirrors on their back so that a creature of ill will would not sneak up and attack them. Figure 2.8 illustrates this effect. An upright arrow, shown on the right, is the object and the inverted image is shown on the left. The object was placed further from the mirror than the focal length.

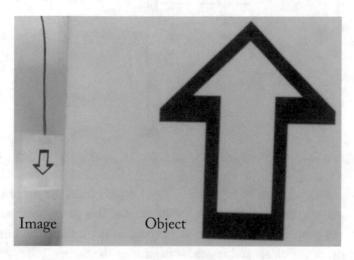

Figure 2.8: Image of an upright arrow inverted by a concave mirror. The image is real, inverted, and reduced in size.

Calculating the Image Distance for a Concave Mirror ($d_0 < f$)

Now we will move the object closer to the mirror. In this example we place the object at a distance smaller than the focal length. This arrangement is shown below in Figure 2.9. What type of image do we expect? Before we trace the rays let's think about the image produced by a plane mirror. In a way, the image produced by the configuration shown in Figure 2.9 shares something in common with the plane mirror. Remember that when you look at your image in a plane mirror it seems to be coming from behind the mirror. We previously mentioned that such an image is called virtual. When a real image is formed you can place a paper screen at the image location and see the image projected onto the paper. If you tried that with a virtual image, the paper screen would seem to be located inside the mirror. Of course, you cannot actually do that. Our brain interprets the image as having originated behind the mirror. As far as our visual perception is concerned the reflection that changed the directions of the rays does not exist. This is why we are so easily fooled or confused by mirrors. We have to tell ourselves that the image is not a real object but merely a reflection.. Think of that when you are in a small room that uses mirrors to make it feel larger.

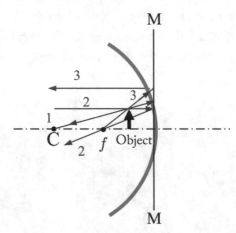

Figure 2.9: Ray diagram for a concave mirror and an object closer that the focal length ($d_o < f$).

For the configuration shown in Figure 2.9, Ray 1 travels from the tip of the object along a line that extends from the center of curvature to the mirror. This time the object is placed closer to the mirror than the position of the center, so we have to extrapolate the line from the center of curvature to the tip of the object and continue the line to the mirror. The ray then reflects back down the line passing back through the center of curvature. Ray 2 originates at the tip of the object and travels along a line that is directed parallel to the optical axis. This ray is reflected off the mirror and passes through the focal point. Thus far the two rays that we have drawn do not intersect. What happens when we add a third ray? We now direct Ray 3 along a line that starts at the focal point, intersects the tip of the object and strikes the mirror. Since this ray lies along a line that goes through the focal point, the reflected ray must be parallel to the optical axis. Note that the rays are extended back to the line marked "M" which represents the plane of the mirror and not the arc that represents the mirror's surface. This is the usual convention in ray tracing.

Unlike the previous example, the rays do not seem to converge to a point on the same side of the mirror from which they originate. If we extrapolate the reflected rays back through the mirror, as shown by the dashed lines, they do converge to a point which is behind the mirror as illustrated in Figure 2.10.

The ray diagram tells us that the image seems to come from behind the surface of the mirror so it is a virtual image.

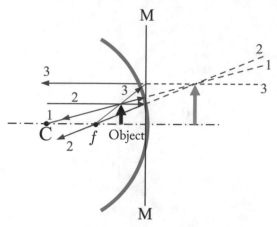

Figure 2.10: Ray diagram for a concave mirror that produces a virtual image.

Now we can examine the mathematics of this configuration. Recall our two mathematical rules for using the mirror equations.

- **Rule 1:** The focal length f is positive if the focal point is on the same side of the mirror as the incident ray and it is negative if it is on the opposite side. This means that for a concave mirror the focal length is positive and for a convex mirror it is negative.

- **Rule 2:** The object and image distances (d_o, d_i) are positive if they are on the same side of the mirror as the incident light ray. If they are on opposite sides of the mirror, then they are negative. A negative image distance indicates that the image is virtual and a positive value indicates that it is a real image.

From these two rules we have a positive focal length and expect a negative image distance. A less cluttered version of the diagram is shown in Figure 2.11, which gives the relevant mathematical parameters.

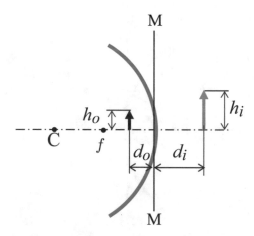

Figure 2.11: Method for calculating the image distance and height for a concave mirror. The object is placed at a distance less than the focal length.

Now we can apply the mirror equation

$$\frac{1}{f} = \frac{1}{d_o} + \frac{1}{d_i},$$

with f = 10.0 cm and d_o = 5.00 cm. We then have

$$\frac{1}{10.0\,cm} = \frac{1}{5.00\,cm} + \frac{1}{d_i}.$$

Solving for the image distance we obtain d_i = −10.0 cm. The distance is negative which indicates a virtual image, just as we expected. We can use the distance information to calculate the magnification and the height of the image:

$$m = -\frac{d_i}{d_o} = -\frac{-10.0\,cm}{5.00\,cm} = 2.00 = \frac{h_i}{h_o}.$$

The height of the object is h_i = 4.00 cm so the height of the image is m × h_o = h_i = 2.00 × 4.00 cm = 8.00 cm. The image is upright and larger than the object as indicated by the ray diagram. Figure 2.12 shows an actual image of a meter stick produced by a concave mirror with the object placed closer that the focal length. This is the second image mentioned by the 墨子 Mòzǐ. Recall that the text states that the explanation lies in whether the object is outside or inside the center of curvature. Perhaps a better wording would be inside or outside the focal length.

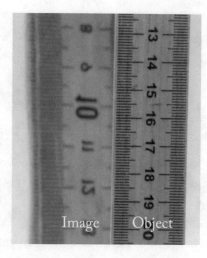

Figure 2.12: A meter stick placed at a position less than the focal length of a concave mirror. The image is virtual, upright, and enlarged.

Calculating the Image Distance for a Concave Mirror ($d_0 = f$ and $d_0 = \infty$)

What happens if the object is placed at the focal point ($d_o = f$)? In that case the equation becomes

$$\frac{1}{f} = \frac{1}{f} + \frac{1}{d_i} ,$$

which yields $\dfrac{1}{f} - \dfrac{1}{f} = 0 = \dfrac{1}{d_i}$.

This means that the image distance goes to infinity ($d_i = \infty$). If you walk toward a concave mirror and look at your own reflection, you will come to a point at which the image becomes all distorted and unclear. When that occurs you are standing at the focal point. This concept is illustrated in Figure 2.13 which shows a meter stick placed very close to the focal point of a concave mirror. The image is highly distorted since the image distance approaches infinity.

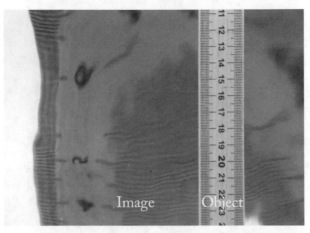

Figure 2.13: A meter stick and its image when placed near the focal point of a concave mirror.

What about an object infinitely far away? Light from the Sun is a very good approximation of a source infinitely far away. If we substitute $d_o = \infty$ then the term containing the object distance becomes

$$\frac{1}{d_o} = \frac{1}{\infty} = 0 \,.$$

With this substitution the mirror equation becomes

$$\frac{1}{f} = \frac{1}{\infty} + \frac{1}{d_i} \,.$$

This tells us that $\frac{1}{f} = \frac{1}{d_i}$, or $f = d_i$. Thus, the Sun's ray form a bright spot at the focal point. If you wish to experimentally determine the focal length of a mirror, simply measure the distance at which the Sun's rays converge to a bright spot. Be careful, this is a great way to start a fire. In fact that is just what people did with "burning mirrors." Directing the rays of the Sun with a bronze mirror was actually an important part of ancient Chinese ritual.

Convex mirrors that were used to start fires are often called 阳燧 yáng suì. The character 阳 yáng is the male or positive principle in the Daoist philosophical idea of 阴阳 yīnyáng . The right side of the character 阳 yáng is the radical 日 rì which means the Sun. 燧 suì is a general term for fire starting. This can apply to fires started by methods such as flint, a friction fire drill, or the rays of the Sun. The left side of the character 燧 suì contains the radical 火 huǒ which means fire. The character looks somewhat like a flame.

Convex Mirrors Described in the 墨子 Mòzǐ.

Just as the 墨子 Mòzǐ describes the properties of concave mirrors it also tells us something about convex mirrors. This description is covered in **B24.**

> In a convex mirror the image is in one case smaller and in one case larger yet necessarily upright. The explanation lies in what is appropriate.
>
> When the object is near, then what is reflected is large and the image is also large. When it is distant, then what is reflected is small and the image is also small and necessarily upright.

I have left out the last line of the translation since there is some doubt about the last few characters. Our main objective is to see if this ancient description accurately describes what we can discern by ray tracing and calculation. Figure 2.14 shows the ray diagram for an object placed some distance away from a convex mirror. Ray 1 originates at the tip of the object and is directed along a line that extends from the object through the center of curvature. This ray must reflect back on itself. Ray 2 is directed parallel to the optical axis. This ray intersects the plane of the mirror at the line M-M and is reflected back along a line that is directed through the focal point. Ray 3 originates at the tip of the object and travels along a line that passes through the focal point. This ray reflects back parallel to the optical axis.

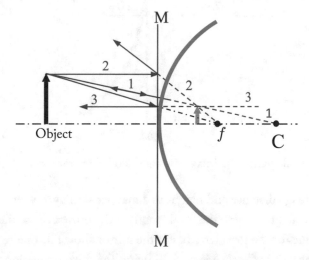

Figure 2.14: Ray diagram of an object placed in front of a convex mirror

None of the rays converge on the same side of the mirror as the object. If we extrapolate the reflected rays in straight line back through the surface of the mirror, they do intersect. Since the intersection point is behind the mirror the image is virtual. We also observe that the image is reduced in size. You may have noticed this effect when look at the side view mirror of an automobile. The mirror on the passenger side of the vehicle is often a convex mirror which affords the driver a wide viewing angle. Such mirrors are marked with a warning which reminds the viewer that "Objects in the mirror are closer than they appear." We tend to judge distance by size. When we see something reduced in size, compared to what our experience tells us is the objects true size, we interpret the object as being far away. The reduced image size produced by a convex mirror, gives the impression that the object is further away than it actually is. This could lead the driver to erroneously conclude that the adjacent vehicle is at a safe distance and result in an accident.

Calculating the Image Distance for a Convex Mirror.

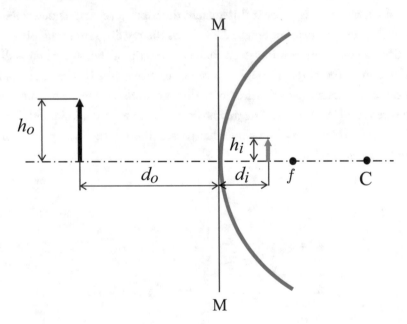

Figure 2.15: Method for calculating the image distance and height for a convex mirror.

We can calculate the image distance and height in a manner similar to what we have already done. We do have to pay special attention to the focal length of the mirror. Since the mirror is convex, we must use a negative value for the focal length. For the mirror shown in Figure 2.15 the focal length is $f = -15.0$ cm and the object distance is $d_o = 28.0$ cm. The mirror equation gives

$$\frac{1}{f} - \frac{1}{d_o} = \frac{1}{d_i} = -\frac{1}{15.0\,cm} - \frac{1}{28.0\,cm},$$

which yields $d_i = -9.77$ cm. The image distance is negative, indicating a virtual image just as we expected.

We can calculate the magnification of this mirror from

$$m = -\frac{d_i}{d_o} = \frac{-(-9.77\,cm)}{28.0\,cm} = 3.49 \times 10^{-1} = \frac{h_i}{h_o}.$$

For an object height of h_o = 12.5 cm, we obtain an image height of h_i = 4.36 cm. This indicates that the image is reduced and upright as we expected. Figure 2.16 shows the image of a meter stick produced by a convex mirror. The image is virtual, upright, and reduced in size.

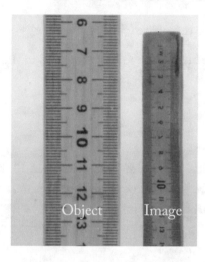

Figure 2.16: A meter stick and its image formed by a convex mirror. The image is virtual, upright, and reduced in size.

Unlike the concave mirrors, convex mirrors will always produce an upright image. We can see this result mathematically if we examine the form of the equations for the image distance and the magnification. First, we will explore the image distance equation

$$\frac{1}{f} - \frac{1}{d_o} = \frac{1}{d_i} = -\frac{1}{15.0\,cm} - \frac{1}{d_o}.$$

As the object distance increases the $\frac{1}{d_o}$ term decreases and for an infinite object distance we have

$$\frac{1}{d_o} = \frac{1}{\infty} = 0 \text{ and } \frac{1}{d_i} = -\frac{1}{15.0\,cm}.$$

This means that the image distance d_i would asymptotically approach d_i = -15.0 cm. Since the magnification is given by

$$m = -\frac{d_i}{d_o}.$$

We would obtain $m = -\dfrac{-15}{\infty} = 0$. That is the magnification would asymptotically approach zero.

We should also consider the case of a very small object distance compared to the focal length. In the equation

$$\frac{1}{f} - \frac{1}{d_o} = \frac{1}{d_i},$$

we see that as the object distance becomes very small, the $-\dfrac{1}{d_o}$ term grows to be very large and we can ignore the $\dfrac{1}{f}$ term. We then have $-\dfrac{1}{d_o} \approx \dfrac{1}{d_i}$ and $d_o \approx -d_i$. This means that the image appears to be the same distance behind the mirror compared to the distance of the object in front of the mirror.

The magnification is then

$$m \approx -\frac{-d_0}{d_o} \approx 1.$$

Both of these mathematical results are clearly shown in Figure 2.17. Referring back to the passage in the 墨子 Mòzǐ we are told that when the object is near the image is large and when it is far the image is small. In both cases the images is upright.

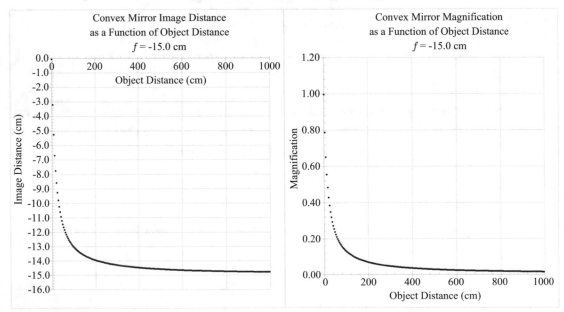

Figure 2.17: Behavior of the image distance and magnification as a function of distance for a convex mirror.

The nonlinear behavior of the magnification leads to some interesting distortions of the image produced by a convex mirror. There is an exaggerated feeling of perspective with is clearly show in Figure 2.18.

This image shows the distortion of the nice horizontal lines of a brick wall and how the magnification falls off quickly with distance.

Figure 2.18: Exaggerated perspective caused by the nonlinear magnification as a function of distance.

2.3   SPHERICAL ABERRATION

Spherical mirrors are relatively easy to make but they do suffer from an optical distortion known as spherical aberration which is illustrated in Figure 2.19.

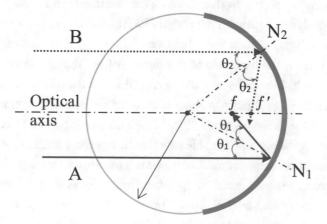

Figure 2.19: Illustration of spherical aberration. The two incident rays are parallel to the optical axis but do not focus at the same point.

Rays A and B are both parallel to the optical axis and ideally we would hope that they would both pass through the focal point f. Ray B strikes the mirror further away from the center line than ray A. We represent the line normal to the mirror at these two points as N1 and N2. By the law of reflection we expect the incident and reflected angles to be the same. However, the angles θ_1 and θ_2 are not the same. Ray B crosses the optical axis at the point marked f', which is slightly behind the focal point f. Rays striking near the edges of the mirror would cross the axis even further behind the focal point f. The result is that rays striking the mirror at different distance from the center form a line of multiple points along the optical axis. This makes the image blurry. There are two ways to minimize this effect. The first is to change the shape from spherical to parabolic. Unfortunately, a parabolic shape is much more difficult to grind. The other method is to ues a mirror whose diameter is small compared to its radius of curvature. It is also possible to obstruct the light striking the mirror so that rays only reflect near the center. Placing an aperture in front of the mirror helps to restrict the light rays to the central portion of the mirror.

2.3.1   THE 魔镜 MÓJÌNG OR MAGIC MIRROR

The mirror shown below in Figure 2.20 is known as a diaphanous mirror. Diaphanous or δĭἄφᾰνής is a Greek word that means transparent, clear, or translucent. The formal Chinese name for such a mirror is 透光镜 tòuguāngjìng or light penetrating mirror. The character 透 tòu means to penetrate or pass through. 光 guāng is the character for light and has something like rays at the top of the character. More colloquially they are known as 魔镜 mójìng or magic mirrors. The character 魔 has several meanings including devil and magic. The mirror does not look transparent does it? It looks like a solid disk of bronze, but it does have some strange properties.

透光镜 tòuguāngjìng were introduced during the Western Han Dynasty 西汉 Xīhàn (Murry and Cahill, 1987). This dynasty spanned the period 206 BC–9 AD and is part of a period that most historians describe as China's golden age. Many examples of 透光镜 tòuguāngjìng are decorated with Chinese characters on the back side of the mirror. When the mirror reflects bright sunlight the characters on the back side can be seen in the reflection. So the mirror is not transparent in the sense that you can see through it but somehow what is on the back side it transmitted through to the reflecting surface. Our hero polymath 沈括 Shěn Kuò (see Chapter 1) mentions this type of mirror in work 梦溪笔谈 Mèng Xī Bǐtán. He says that he has three ancient mirrors, but among his own mirrors and those of his acquaintances only one is a transparent mirror. 沈括 Shěn Kuò retells the theory that the pattern is transferred to the reflective surface as slight distortions. These distortions were thought to be caused by a difference in the cooling rate of the metal when the mirror was cast. People believed that the areas in which a character stood out in relief on the back surface would necessarily be thicker and cool at a different rate that the thinner parts. These distortions produced a latent image that could not be seen with the naked eye, yet did reflect sunlight slightly differently than the surrounding area. 沈括 Shěn Kuo says that this seems to be a reasonable theory and admits that ancient artisans must have had a special method to make such mirrors. This seems to imply that in his day this was considered a lost art. The 透光镜 tòuguāngjìng shown in Figure 2.20 was made for the author by Mr. 董子俊 (Dong Zijun). Mr. Dong is the director of the Bronze Casting Technology Research Institute of the Zhouyuan Museum (周原博物馆青铜范铸工艺研究所) in Shaanxi province (陕西省 Shǎnxī Shěng).

Figure 2.20: A 透光镜 tòuguāngjìng or light-penetrating mirror.

The 透光镜 tòuguāngjìng shown here would have been even more mysterious to 沈括 Shěn Kuò. This mirror, from the author's collection, produces pattern that is totally unrelated in shape to the symbols on the back side. As shown in Figure 2.21, the back side of the mirror has a lion and a dragon while the reflective side casts an image of the character 福 fú. So much for the differential cooling theory. Perhaps that was the technique in ancient times. The character 福 fú can mean good fortune, happiness, and luck. Our eyes have tremendous dynamic range and the contrast differences that we can see with our eyes is much better than what can be reproduced in a book. For this reason we have enhanced the reflected image to make the character more obvious. The lower enhanced image is black and white and has the appearance of a traditional Chinese stone rubbing. We should point out that the reflected character is written in calligraphic style so it look slightly different from the printed version 福. A more artistic rendering of the character 福 fú is shown in Figure 2.22. This is a decorative paper cut that is commonly used at Lunar New Year. You may notice that the character is shown upside down and right-side up. Like everything else in Chinese culture, there is a long story behind that. The Chinese language employs a great deal of word play based on homophones. If something is placed upside down it is described as 倒 dào. However, the character 到 dào can mean to arrive. So the statement "福倒了 fú dào le" can mean your 福 fú is upside down, but it also sounds like you are saying the good fortune has arrived.

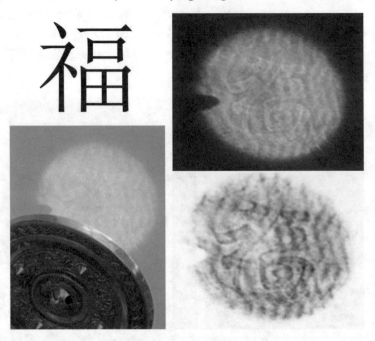

Figure 2.21: The character 福 fú or good fortune produced by a magic mirror along with various enhancements of the image.

Figure 2.22: The character 福 fú depicted in a paper cut used for decoration. On the right side the 福 fú is upside down or 倒 dào which is a play on words.

You may be wondering about those birds. Do they have a symbolic meaning? Of course, its Chinese culture and everything has a symbolic meaning. So do the animals on the mirror. The birds are magpies or 喜鹊 xǐquè. They are a very important symbol in Chinese culture and found in several stories and holidays. The 喜 xǐ character in their name has meanings such as like, love, and enjoy. Magpies play a special role in the story of the Cowherd and Weaver Girl known in Chinese as 牛郎织女 niúlángzhīnǔ. On the seventh day of the seventh month, reckoned by the lunar calendar, a flock of magpies forms a celestial bridge across the Milky Way. This special day is called 七夕 Qīxī or the double seven festival. The character 七 qī means seven and 夕 xī means evening. Sometimes this festival is called Chinese Valentine's Day. Tragically, this is the only time of the year when the two lovers are allowed to be together. Otherwise, they are separated by the 银河 Yínhé, which is literally the silver river and symbolically represents the Milky Way. 喜鹊 xǐquè are often use to decorate mirrors and have another special purpose. In one ancient story, a husband and wife break a mirror into two pieces. Then, if one partner is unfaithful, the mirror turns into a 喜鹊 xǐquè to alert the other partner.

As for those four nipples or protrusions on the back of the mirror, they were a common Han Dynasty symbol. Notice they are arranged to form the corners of a square. Square shapes represent the Earth the four cardinal directions and the mysterious 四象 sìxiàng of Chinese astronomy/astrology. These creatures are 青龙 Qīnglong the Azure Dragon of the East, 白虎 Báihǔ the White Tiger of the West, 朱雀 Zhūquè the Vermilion Bird of the South, and 玄武 Xuánwǔ the Black Tortoise of the North. Chinese astronomy and astrology is intensely complex and would take an entire book to explain. Perhaps it will be the next book in this series.

REFERENCES

Cammann, S. (1953). The Lion and Grape Patterns on Chinese Bronze Mirrors. *Artibus Asiae*, 16(4), 265–291. DOI: 10.2307/3248647. 57

Graham, A. (1978). *Later Mohist Logic, Ethics and Science*. The Chinese University Press. 59

Johnston, I. (2010). *The Mozi, A Complete Translation*. (I. Johnston, Trans.) Columbia University Press. 59, 60

Murry, J. K. and Cahill, S. E. (1987). Recent Advances in Understanding the Mystery of Ancient Chinese "Magic Mirrors." *Chinese Science*, 8, 1-8.

Needham, J. and Wang, L. (1956). *Science and Civilisation in China, History of Scientific Thought* (Vol. 2). Cambridge University Press. 59

Zhu, S. and He, T. (1993). Studies of Ancient Chinese Mirrors and Other Bronze Artefacts. *Metal Plating and Patination*, 50-62. DOI: 10.1016/B978-0-7506-1611-9.50009-1. 59

CHAPTER 3

Metals 金属 Jīnshǔ

3.1 BITTER WATER IN 梦溪笔谈 MÈNG XĪ BǏTÁN

In this chapter we once again take as our starting point a quote from our polymath hero 沈括 Shěn Kuò. In his work 梦溪笔谈 Mèng Xī Bǐtán he describes a rather amazing bit of what I like to call proto-chemistry. Others may regard this as simply alchemy but so much of what alchemists discovered can be explained by simple chemical principles that were unknown at the time. Their experiments and observations did help modern scientists formulate early chemical principles. In Section 25 (卷 juàn 二十五), the sixth paragraph we read the following remarkable description.

> In the Xìnzhōu district of Yánshān County there is a bitter spring that flows into a mountain stream. When water scooped out from this stream is boiled off it produces dǎn fán crystals. If you heat the dǎn fán it turns into copper. If the dǎn fán containing water is allowed to simmer for a while in an iron kettle it transforms into copper. Thus water is transformed into copper. This transformation of matter is a deep and unfathomable mystery.

> According the Basic Questions section of the Inner Canon of the Yellow Emperor, there are five phases of matter in the sky and five phases on the Earth. When the qì of the Earth is in the sky it becomes damp. Earth gives rise to metal and stone, damp also gives rise metal and stone. This proves that the ideas of the Inner Cannon are correct. Also, water dripping from the roof of a cave can form stalactites… This is the nature of the five phases.

You will find many variations on this translation and I have tried to render it with a form that helps clarify the chemistry. I have only given a translation of the first part of ancient text, but it is enough to establish the idea. Throughout this book we have established a reoccurring theme. First, we start with an ancient Chinese discovery or technology. We explore how this discovery or technology was understood in its cultural and historical context. Then, we try to find the physics or chemistry hiding in the background. Finally, we subject the technology to some form of experimental testing. We can apply this same theme to the statement concerning this nearly magical production of copper. Before we dive into the text we are going to take a journey through chemistry, materials science, metallurgy, and Chinese numismatics. Then we will be prepared to see all the scientific and cultural connections surrounding what 沈括 Shěn Kuò is telling us. In the discussion that follows, copper will play an important role. As we will see, copper is the main ingredient in the alloy that changed history. Chapter 5 of the experimental section also contains several activities aimed and methods of extracting copper.

3.2   BRONZE CASTING

When I teach these topics at my home institution, Mercer University, I take bronze casting as the starting point. The ancient Chinese were masters at casting bronze and their works of great beauty can be found in museums all over the world. They achieved a level of technical mastery and beauty not found in any other cultures that I am aware of. Bronze casting is used as a way of introducing concepts such as metals, alloys, the periodic table and crystal structures. One of the goals in writing this book was to transfer knowledge and techniques to others who might be interested in incorporating these ideas into their own teaching method. Bronze casting is not something that is easily incorporated into the typical classroom. I have been casting metals for some time and have a great deal of specialized equipment that enable my students to safely engage in this activity. Some of their handiwork is shown in Figure 3.1.

Figure 3.1: Examples of cast bronze objects made by students. The wax model and the cast bronze piece is shown for each.

The method we employ in lab is known as lost wax casting. One first makes a wax model of the desired object. The wax is surrounded by plaster and then melted out leaving a cavity that faithfully reproduces the shape of the object. Molten bronze is then poured into the cavity to create a work of art.

Figure 3.2: Pouring molten bronze in the lab.

3.3   COPPER AND TIN ORE

In order to make bronze you need at least two materials—copper 铜 tóng and tin 锡 xī. Figure 3.3 shows copper and tin shot in a ratio used to make bronze. Lead and other elements can be added to modify the material properties. Both copper (Cu) and tin (Sn) are soft materials but an amazing transformation takes place when they form the bronze alloy. In this chapter we will discuss some basic materials science and take a deeper look at the chemistry involved. We normally think of tin as a nice shiny white metal and it is in its elemental form. In nature it looks quite different. Shown in Figure 3.4 is the mineral Cassiterite 锡石 xīshí which is how you are more likely to find tin. It looks very different from the white metal shown above. The Cassiterite shown is an oxide of tin (SnO_2). In order to get the nice white metallic form we must reduce the oxide.

Copper is found all over the world and many ancient cultures learned how to fashion it into useful objects. Occasionally copper can be found in its native form. By native form we mean in a nearly pure or uncombined state.

Sometimes native copper is found in the form of nuggets or even thin plates. Before people learned how to cast copper and combine it with other metals, they could shape native copper into useful forms. A thin sheet could become a knife blade or some other type of tool. Currently, the earliest use of native copper in China can be traced back to the late 4th or early 3rd century BC (Mei, 2000).

Figure 3.3: Copper (Cu) and tin (Sn) in a typical ratio (80:20 w/w) used to produce bronze. The white metal is the tin.

Figure 3.4: The mineral Cassiterite (SnO_2) 锡石 xīshí which is a common source of tin ore.

Working native copper had its limitations. The hammering, bending, and grinding needed to shape the metal work hardened the material making it more difficult to bend and causing it to become brittle. Melting copper takes additional technology since it melts at 1085°C slightly higher than gold but less than iron. Native copper is a very limited resource and if you are going to use it extensively or combine it in an alloy you need a rich supply. Other than native copper and copper oxides, copper can be mined form ores. There are about a dozen different copper bearing ores. Two common ore types consist of sulfides or carbonates. As the names imply, the copper is combined with other elements. A very common sulfide ore is chalcopyrite $CuFeS_2$, which is shown in Figure 3.6. In this ore the copper is combined with iron and sulfur.

Figure 3.5: A large mass of native copper from the Keweenaw Peninsula, Michigan. Collected by Bill Christy. Courtesy of the Museum of Arts and Sciences, Macon, Georgia.

Figure 3.6: Mineral samples (a) and (b) Chalcopyrite ($CuFeS_2$) a common copper ore. Sample (c) Covellite CuS a rare copper mineral.

If you saw these stones lying on the ground you might think they contain some valuable metal. The brassy to golden-yellow color is sure to attract attention. Chalcopyrite can be smelted and copper can be extracted through a process called roasting. There are numerous processing steps that require the use of other materials such as silica (SiO_2) or limestone ($CaCO_3$) to produce intermediate compounds which are further refined. The processing steps also require high temperatures in the range of 1200°C.

3.4   MALACHITE

Malachite $Cu_2CO_3(OH)_2$ is a copper carbonate and extracting copper from it is relatively simple. If we heat malachite to about 350°C it will produce copper oxide (CuO), carbon dioxide (CO_2), and water (H_2O). The reaction is:

$$Cu_2CO_3(OH)_2(s) \xrightarrow{\triangle} 2CuO(s) + CO_2(g) + H_2O(g).$$

Now we have copper (II) oxide which is reduced to elemental copper (Cu) by heating in the presence of carbon (C) and produces carbon dioxide:

$$2CuO(s) + C(s) \xrightarrow{\triangle} 2Cu(s) + CO_2(g).$$

Figure 3.7: Malachite sample collected by Margaret S. Brewer. Courtesy of the Museum of Arts and Sciences, Macon, Georgia.

Charcoal placed around the malachite serves as the carbon source for reducing the oxide. The entire process can be carried out in a clay container placed in a pottery kiln. Ancient kilns of the Shang Dynasty were capable of producing temperature of about 1,100–1,200°C, which means they could melt copper. It is also possible to achieve high enough temperatures with a campfire, but air must be blown into the fire. Malachite is called 孔雀石 kǒngquè shí in Chinese. The character 石 shí means stone and 孔雀石 kǒngquè is a peacock. One of the greatest archaeological finds of the late 19th century was near the modern city of 安阳 Ānyáng in 河南 Hénán province. Nearby the city is a World Heritage site known as 殷墟 Yīnxū in which we find the ruins of Yīn, one of the ancient capitals of China during the Shang Dynasty 商朝 Shāngcháo (16th century BC to 1046

BC). Extensive excavations of 殷墟 Yīnxū were conducted in the early 20th century and what was found there has a great deal to do with this discussion of copper. Noted archaeologist Cheng Te-K'un (郑德坤 Zhèng dékūn) points out that in 1932 the ruins of a 商朝 Shāngcháo foundry were discovered. A large piece of malachite with a mass of 18.8 kg was found in addition to pottery shards, crucibles, mould (a shaped cavity used to give a definite form to fluid or plastic material), and charcoal (Cheng, 1974). This suggests that malachite was used as copper ore during the period 1350–1046 BC, when 殷 Yīn was the Capital city. Other types of ore may have been used at that time, but smelting malachite was relatively simple.

Copper is important in this history of metallurgy, warfare and technology but it has its limitations. Tools and weapons made of copper were effective in their time but they just did not hold and edge. Copper is soft metal and can be combined with other metals to form an alloy. Alloys are mixtures of metals, and occasionally nonmetals, which produce a final product with properties that can be very different from those of its components. The resulting material still has the characteristics of a metal but it is often engineered to have qualities that are better suited for a particular application than its constituents. We encounter many alloys in our daily life. Do you have any gold jewelry? Pure gold is very soft and easily distorted. For this reason it is common to make jewelry from harder gold alloys that retain their shape. Sterling silver contains 92.5% silver alloyed with copper and other metals. Iron combined with carbon produces the well-known alloy called steel. Adding some chromium and nickel to your steel turns it into stainless steel. All of these metals have some common characteristics.

3.5   PROPERTIES OF METALS

What makes a metal a metal? The exact definition of metal can be complicated. Some nonmetals can become metals under the right conditions. There is a class of materials known as metalloids which have properties in between those of a metal and nonmetal. To make things even more complicated, nonmetals such as iodine become metals under tremendous pressure. Hydrogen is sometimes listed as a metal, nonmetal, or in a class of its own. Normally, we think of it as a gas, but under extreme pressure, such as in the core of Jupiter, it behaves as a metal. Astatine, atomic number 85, is sometimes classified as a metal, nonmetal, or metalloid.

The periodic table shown in Figure 3.8 reveals that most of the known elements are metals. The element names in Figure 3.8 are based on the International Union of Pure and Applied Chemistry (IUPAC) nomenclature as of December 1, 2018. You may see different versions of the periodic table or different classifications for some of the metalloids. But we are interested in what makes a metal a metal.

1 1A	2 2A	3 3B	4 4B	5 5B	6 6B	7 7B	8 8B	9 8B	10 8B	11 1B	12 2B	13 3A	14 4A	15 5A	16 6A	17 7A	18 8A
1 H																	2 He
3 Li	4 Be											5 B	6 C	7 N	8 O	9 F	10 Ne
11 Na	12 Mg											13 Al	14 Si	15 P	16 S	17 Cl	18 Ar
19 K	20 Ca	21 Sc	22 Ti	23 V	24 Cr	25 Mn	26 Fe	27 Co	28 Ni	29 Cu	30 Zn	31 Ga	32 Ge	33 As	34 Se	35 Br	36 Kr
37 Rb	38 Sr	39 Y	40 Zr	41 Nb	42 Mo	43 Tc	44 Ru	45 Rh	46 Pd	47 Ag	48 Cd	49 In	50 Sn	51 Sb	52 Te	53 I	54 Xe
55 Cs	56 Ba	71 Lu	72 Hf	73 Ta	74 W	75 Re	76 Os	77 Ir	78 Pt	79 Au	80 Hg	81 Tl	82 Pb	83 Bi	84 Po	85 At	86 Rn
87 Fr	88 Ra	103 Lr	104 Rf	105 Db	106 Sg	107 Bh	108 Hs	109 Mt	110 Ds	111 Rg	112 Cn	113 Nh	114 Fl	115 Mc	116 Lv	117 Ts	118 Og

57 La	58 Ce	59 Pr	60 Nd	61 Pm	62 Sm	63 Eu	64 Gd	65 Tb	66 Dy	67 Ho	68 Er	69 Tm	70 Yb
89 Ac	90 Th	91 Pa	92 U	93 Np	94 Pu	95 Am	96 Cm	97 Bk	98 Cf	99 Es	100 Fm	101 Md	102 No

Metal | Metalloid | Nonmetal | Unknown

Figure 3.8: Periodic table of the elements reflecting IUPAC nomenclature adopted December 1, 2018.

Think about the common metals you know. What are some of their characteristics? Metals are often shiny, opaque, conduct electricity, and conduct heat. In terms of mechanical properties, they can be drawn into wires which is called ductility. Metals can be hammered into thin sheets, which is a property called malleability. Gold, silver, and copper leaf are very thin sheets of metal that are adhered to nonmetals often for artistic purposes. That is, they are malleable; you can make a thin continuous sheet by striking with a mallet. All these properties are the result of the behavior of the electrons in the metal. Like most things in chemistry it all depends on the behavior of the electrons. Atoms can form bonds and the bonds formed in a metal are of a different nature than covalent and ionic bonds. A common characteristic of metals is that they have comparatively lower ionization energies. That is, it take less energy to give up a valance electron. Copper has a first ionization energy of 745.5 kJ/mol compared to Fluorine's 1681 kJ/mol and Helium's 2372 kJ/mol.

Early models of atoms often depicted the electron as orbiting a central nucleus much like a planet orbiting around the sun. As our understanding progressed, we saw the motion of the electron more like a cloud but still orbiting about a particular nucleus. Quantum mechanics helped us to understand the complex shapes of these electron orbitals. Another characteristic of metals is that thy have unfilled orbitals and only a few valance electrons. Valance electrons reside in the outermost shells and are the ones that interact the most with other atoms. Inner electrons usually do not contribute in bonding. Copper is interesting in terms of its valance electrons in that it can have two different oxidation states. The electron configuration of copper is $1s^2\,2s^2\,2p^6\,3s^2\,3p^6\,3d^{10}$ $4s^1$. You may remember from high school chemistry that the 4s electron state fills before the 3d. Copper can lose one electron from the 4s level or one from the 4s and one from the 3d level. This creates two possible oxidation states Cu (I) and Cu(II). This is not unique to copper but occurs for other transition metals of the B-group. Columns in the periodic table tells us something important about atoms. The elements in the same column have similar chemical characteristics because they have similar outer electron configurations.

When we bring atoms together to form a piece of metallic material, such as a lump of copper, something interesting happens to the valance electrons. Instead of identifying each valance electron with a particular atom, the electrons delocalize. They are free to migrate around throughout the material. The positive metal ions are now immersed in a sea of electrons. The electrostatic attraction between the ions and the sea of electrons holds the material together but the individual electrons are free to move round in the material. This gives rise to some of the properties we just mentioned. Malleability and ductility are direct physical manifestations of metallic bonding. Layers of metal atoms can freely slide past each other as a metal is deformed. Local bonds around the metal ion are easily broken and reformed so the metal is held together overall, but on an atomic level metal ions can easily readjust their positions. The fact that electrons can move relatively freely in a metal compared to an insulator gives rise to electrical conductivity. The electrons that conduct electricity are not valance electrons but rather conduction electrons and they differ slightly in energy. Without going to deep into solid-state physics, there is an energy gap between valance and conduction electrons. In a metal this gap can be very small or the bands may even overlap. Thus, very little energy is required to promote an electron from the valance band to the conduction band. Electrons in the conduction band are free to move through the material and conduct electricity. Occasionally they have collisions, so the material will not be a perfect conductor and there will be some electrical resistance. Insulating materials have a much higher band gap and more energy is required to move an electron from the valance band to the conduction band. The difference in energy between the valance band and the empty conduction band is called the Energy gap E_g, as shown in Figure 3.9.

Metals are often good reflectors of light which makes them shiny. Reflectivity is also the result of electrons. We can think of light as an electromagnetic wave meaning that it has an electric field component and a magnetic field component. These components vary in time and space. We know that an electric field can exert a force on a charged particle such as an electron. So the electrons are vibrated by the electric field of the incident wave. When an electron, or any charged particle, is accelerated it can produce an electromagnetic wave. This means that vibrating the electrons in the metal produces an electromagnetic wave which is now radiated from the surface of the metal. The electrons cannot respond to any arbitrarily fast vibration. There is an upper limit to the frequency they can respond to. For most metals this upper limit corresponds to frequencies in the ultraviolet portion of the spectrum.

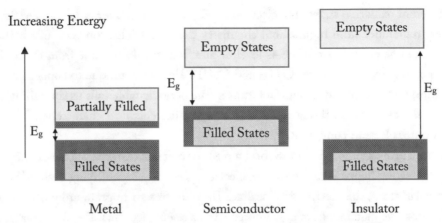

Figure 3.9: Energy band structure for a metal semiconductor and insulator at 0K

For lower frequencies, in the visible portion of the spectrum, the electrons can respond and light is reflected. If all colors in the visible are reflected equally well, then white light falling on the metal reflects back all of the same colors striking the surface. Thus, the metal looks white. If some colors are reflected with a lower intensity, this shifts the color balance of the reflected light. Thus, the light reflected from the metal takes on a different color or hue. Again, it is all about the electrons.

Metals are good conductors of heat. The thermal conductivity of a metal like silver is very high. If you place a plastic spoon and a silver spoon into a cup of tea which spoon feels hottest when you touch it? Silver has a very high thermal conductivity so the end you touch will be closer to the temperature of the hot tea. Plastic has a low thermal conductivity it is a better thermal insulator. That means the end you touch will be at a much lower temperature than the hot liquid. Another simple way to demonstrate the thermal conductivity difference of materials is to touch objects that should all be about the same temperature. If you are sitting at a desk, touch the nonmetallic part of the desk top and then touch some metal object on or near the desk. Which feels cooler? If the objects have come to equilibrium with the room, they should have the same temperature. The metal object will feel cooler. This is because it will conduct heat away from your finger better than a wood or plastic desk top. We interpret the heat energy flowing out of our finger and into the metal object as an indication of the object being cold. For low temperatures, heat energy is mostly transported by the electrons that freely move in the material. At higher temperature vibrations of the crystal structure become more important. So what makes a metal a metal? The answer lies in the sea of electrons that move about through the material.

3.6   BACK TO THE BITTER WATER

We have covered a great deal of chemistry and physics and now we are going to return to the ancient text. We will examine the text and dig out the underlying physics and chemistry. In interpreting this text we will also take a short journey through Daoist alchemy. After our discussion of the text we will finish out the chapter with a short discussion of bronze.

After all the chemistry and physics we have discussed thus far, you may have completely forgotten about the Chinese text that was a starting point. So let us take another look at what 沈括 Shěn Kuò reported.

> In the Xìnzhōu district of Yánshān County there is a bitter spring that flows into a mountain stream. When water scooped out from this stream is boiled off it produces dǎn fán crystals. If you heat the dǎn fán it turns into copper. If the dǎn fán containing water is allowed to simmer for a while in an iron kettle it transforms into copper. Thus water is transformed into copper. This transformation of matter is a deep and unfathomable mystery.

> According the *Basic Questions* section of the *Inner Canon of the Yellow Emperor*, there are five phases of matter in the sky and five phases on the Earth. When the qì of the Earth is in the sky it becomes damp. Earth gives rise to metal and stone, damp also gives rise metal and stone. This proves that the ideas of the *Inner Cannon* are correct. Also, water dripping from the roof of a cave can form stalactites… This is the nature of the five phases.

We will examine his observation piece by piece. I think it is significant to examine the location involved. 铅山县 Yánshān county is in 江西 Jiāngxī province which is located in southeastern China. The area is mountainous and known for its rich mineral deposits. In this very area we can find one of China's major copper mines, the 永平 Yǒngpíng mine. This mine is part of China's main copper producer Jiangxi Copper 江西铜业 Jiāngxītóngyè. Metals of antiquity such as gold, silver, and copper are found in 江西 Jiāngxī province along with more exotic metals such as niobium, uranium and thorium. By metals of antiquity we mean the metals that mankind has used for most of our history. These metals are: gold, silver, copper, tin, lead, iron, and mercury.

Some translations render the name of the county as Qiānshān and sometimes as Yánshān. The Chinese character 铅 yán has both pronunciations but the meaning is Lead, as in the element Pb. In mineralogy, we often group minerals with similar properties. The copper group consists of copper, gold, lead and silver. The comment that the water was bitter is very important. In ancient times, copper miners used smell and taste to help them identify particular minerals. The bitter taste was well known to copper miners particular those involved with the wet copper method. Golas points out that miners engaged in the wet copper process even used variations in taste to determine the copper concentration of the water (Golas, 1999).

Table 3.1: Metals of antiquity used by man for thousands of years before the modern era	
Metals of Antiquity	
Metal	**Melting Point (°C)**
Iron (Fe) 铁 tiě	1,538
Copper (Cu) 铜 tóng	1,085
Gold (Au) 金 jīn	1,064
Silver (Ag) 银 yín	962
Lead (Pb) 铅 qiān	327
Tin (Sn) 锡 xī	232
Mercury (Hg) 汞 gǒng	-39

In this translation I have left 胆矾 dǎn fán untranslated. The term is sometimes rendered as "gall alum," "blue vitriol," or copper sulfate. The term gall is a direct reference to the gallbladder and the sour taste of the bile produced by the liver. It is worth pointing out that animal bile has been used in traditional Chinese medicine for treating a wide range of medical problems including diseases of the skin, eyes, ears nose, throat, digestion, and gynecological disorders (Wang and Carey, 2014). Alum generally refers to hydrated sulfate salts of aluminum such as potassium (K) alum, which has the chemical formula $KAl(SO_4)_2·12H_2O$. Alums are often characterized by their astringent properties and acidic taste. Blue vitriol is a somewhat better term to use but it is an archaic term not likely to be familiar to students. A vitriol is a sulfate and the name is derived from the Latin "vitriolum" which means glassy. It is a description of the sulfate crystals that look like pieces of colored glass. A large piece of Copper (II) Sulfate is shown in Figure 3.10. When water containing 胆矾 dǎn fán is boiled off or evaporated it leaves behind beautiful blue crystals that do look like colored glass.

The blue crystals are the pentahydrate form, $CuSO4·5H2O$. As they are heated, the crystals dehydrate. In its anhydrous form, Copper (II) sulfate is white. Further heating above about 650°C causes the sulfate to decompose and form copper oxide. This process is shown in Figure 3.11.

Figure 3.10: Copper (II) Sulfate Pentahydrate $CuSO_4 \cdot 5H_2O$.

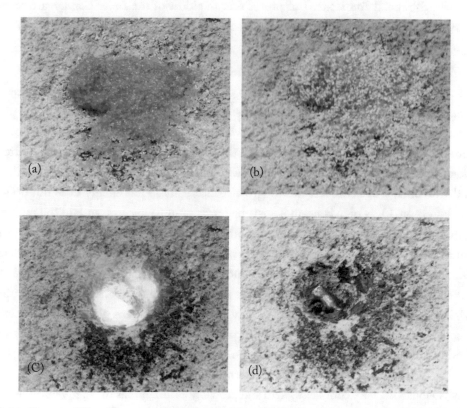

Figure 3.11: Copper (II) sulfate pentahydrate ($CuSO_4 \cdot 5H_2O$) heated with a torch. (a) Before heating, (b) dehydrating, (c) decomposed to molten copper, and (d) oxidized lump of copper at room temperature.

3.7   CONCEPTS FROM CHINESE ALCHEMY

So far we can see that Shěn Kuò's description seems accurate. Now let's look at the next part which seems mysterious.

> If the dǎn fán containing water is allowed to simmer for a while in an iron kettle it transforms into copper. Thus water is transformed into copper. This transformation of matter is a deep and unfathomable mystery.

Heating the water is not actually necessary, the reaction will occur spontaneously at room temperature. Keep in mind that this is not about the iron metal in the kettle transforming into copper as some have suggested. A little further on this becomes clear when he writes:

> According the *Basic Questions* section of the *Inner Canon of the Yellow Emperor*, there are five phases of matter in the sky and five phases on the Earth. When the qì of the Earth is in the sky it becomes damp. Earth gives rise to metal and stone, damp also gives rise metal and stone. This proves that the ideas of the *Inner Cannon* are correct. Also, water dripping from the roof of a cave can form stalactites… This is the nature of the five phases.

Notice that he uses another example of water transforming into a mineral when he mentions stalactites. From the stalactite reference it seems clear that the water is undergoing some kind of transformation and not the iron itself. There is also a great deal of Daoist alchemy in the preceding paragraph. Before we dive into the chemistry of what is happening we should examine how the ancient Chinese understood the process. I have left the term 气 qì (traditional form 氣) untranslated. If you look this character up in a dictionary you will find meanings such as: gas, air, smell, to make angry, and vital energy. It is the last meaning that concerns us. Vital energy does not do us much good as a definition. Qì is qì. Many sinologists have given up trying to translate the term and just leave it untranslated. In Daoist thought 气 qì is often described some kind of active agent that permeates the universe. Traditional Chinese medicine, philosophy, martial arts, and what we may call early science are all intimately connect to the concept of 气 qì. Some have even likened it to wave particle duality, but this seems like a stretch. It is a metaphysical term and very hard to pin down. Although it may be called vital energy, it is not the same as the modern physical concept of energy. Those who believe in the concept think it permeates everything and that its flow and balance have physical effects. In Chapter 1, we encountered the concept of 气 qì in terms of 风水 fēngshuǐ. The concept of 气 qì is intimately connected to Daoism 道教 Dàojiào.

道教 Dàojiào is a very complex system and probably as unfamiliar to most Westerners as is the concept of 气 qì. We can gain a little understanding of 气 qì and its relation to the physical world from another ancient text the 淮南子 Huáinánzi. Here we follow the translation by Major, Meyer, Queen, and Roth (An, 2010). The original text is a collection of essays based on discussions held in the court of the Prince of Huainan-刘安 Liú Ān . This compilation of essays is thought to be from around the year 139 BC during the Western Han 西汉 Xīhàn dynasty. The text is often studied in philosophy classes as an example of how to form a perfect state and contains advice on how to rule wisely. The introductory sections contain lessons on the basics of Daoism and Confucianism with the goal of harmonious governance. Section three of the 淮南子 Huáinánzi treats the topic of 天文訓 tiānwénxùn which may be translated as "the patterns of heaven" or "celestial patterns."

> When Heaven and Earth were yet unformed, all was ascending and flying, diving and delving. Thus it was called the Grand Inception. The Grand Inception produced the Nebulous Void. The Nebulous Void produced space-time; space-time produced the original [undifferentiated] 气 qì. A boundary [divided] the original 气 qì. That which was pure and bright spread out to form Heaven; that which was heavy and turbid congealed to form Earth. It is easy for that which is pure and subtle to converge but difficult for the heavy and turbid to congeal. Therefore, Heaven was completed first; Earth was fixed afterward. The conjoined essences of Heaven and Earth produced yin and yang. The suppressive essences of yin and yang caused the four seasons. The scattered essences of the four seasons created the myriad things. The hot 气 qì of accumulated yang produced fire; the essence of fiery 气 qì became the sun. The cold 气 qì of accumulated yin produced water; the essence of watery 气 qì became the moon. The overflowing 气 qì of the essences of the sun and moon made the stars and the planets. To Heaven belong the sun, moon, stars, and planets; to Earth belong waters and floods, dust and soil.

This little detour into Daoist cosmology may seem very out of place in a book about science but this was the world view of the ancient Chinese. As strange as it seems to us, 气 qì was their go to tool used to explain the nature of things. Another tool was the ubiquitous five-element or five-phase theory 五行 wǔxíng. Again, this is a hard term to translate. When Westerners hear "five element" they think of elements as fixed entities such as the elements on the periodic table. This is not at all how the Chinese used the term. I prefer the term phases but alternative translations are agents, movements, processes. In the West we never think of the "elements" as changeable things but as fixed building blocks. In the Chinese or Daoist understanding of nature the 五行 wǔxíng are not static but rather active agents whose influence either declines (overcoming) or grows (generation) with time. This occurs in a cyclic manner sometimes correlated with the seasons of the year. The cycle of generation is known as 相生 xiāngshēng and the cycle of overcoming is called 相克 xiāng kè. Generating 相生 xiāngshēng, interactions are shown in Figure 3.12.

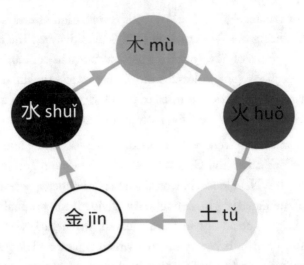

Figure 3.12: The five phases 五行 wǔxíng in the mutual generation sequence 相生 xiāngshēng.

When 沈括 Shěn Kuò tells us that earth gives rise to metal this is what he is talking about. It is not just simply the act of mining metal from ore, but rather a more intimate form of generation. The term 生 shēng in 相生 xiāngshēng can mean birth and they often thought of the Earth phase as giving birth to or creating the Metal phase.

In the original Chinese, the word translated "damp" is 湿 shī which is related to traditional Chinese medicine. According to Traditional Chinese Medicine there are six unhealthy influences that cause illness. These six influences, called 六邪 liùxié, are: excessive wind 风 fēng, cold 寒 hán, heat 暑 shǔ, damp 湿 shī , dryness 燥 zào, and fire 火 huǒ. Traditional Chinese medicine is composed of many different connections and interactions between the 五行 wǔxíng and explaining those details is far beyond what we can discuss here.

The concluding paragraph gets even deeper into Traditional Chinese Medicine. We read:

> According the *Basic Questions* section of the *Inner Canon of the Yellow Emperor*, there are five phases of matter in the sky and five phases on the Earth. When the qì of the Earth is in the sky it becomes damp. Earth gives rise to metal and stone, moisture and damp also gives rise metal and stone. This proves that the ideas of the *Inner Cannon* are correct. Also, water dripping from the roof of a cave can form stalactites… This is the nature of the five phases.

First, we should notice that what happens in the sky or heavens (天 tiān) is mirrored on the earth (土 tǔ). This relationship is a basic principle in many forms of Chinese philosophy. He then points out that damp also gives rise to metal. One of the interactions in Traditional Chinese Medicine is sometimes expressed as 土生湿而金存 tǔ shēng shī ér jīn cún. Which can be translated as:

"The earth element (phase) creates damp and the metal element (phase) stores it." This interaction is shown below in Figure 3.13. It seems likely that this is what 沈括 Shěn Kuò is referencing.

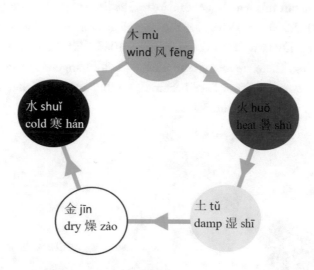

Figure 3.13: Associations of the five phases and the unhealthy influences according to Traditional Chinese Medicine.

At the end of his description, 沈括 Shěn Kuò declares "This is the nature of the five phases." This last line is like putting QED (quod erat demonstrandum) at the end of a mathematical proof or philosophical argument. He is telling us that yes, the transformation of 胆矾 dǎnfán containing water is consistent with our understanding of nature. Not only does it fit into the world view of 沈括 Shěn Kuò, but it also can be explained through our modern understanding of chemistry. Remember, that throughout this text we first look at the traditional Chinese explanation based on their world view. The ideas of traditional Chinese medicine and the interactions among the 六邪 liùxié are still quite common in the modern Chinese community and one hears them all the time in discussions about illness. We will take up the chemical explanation of the bitter water transformation later in the chapter.

3.8 BRONZE

Why are we spending time on this topic and the production of copper? Again, it is the objective of this book to show how ancient Chinese technologies can be used to teach modern day students some physics and chemistry. First, we look at how the ancients understood the technology in their own cultural context and then we see what we can learn from it using modern scientific methods and analysis. We have already mentioned the idea of an alloy. In the ancient world the alloy that changed history was bronze. It is not entirely clear just when the ancient Chinese started to produce bronze. The Neolithic 马家窑文化 mǎjiāyáo wénhuà or mǎjiāyáo culture of may have been the first bronze culture in China but this is a subject of debate. The time period of the 马家窑文化 mǎjiāyáo wénhuà is about 3,300–2,000 BC. Whatever the origins of bronze in China, by the time of the 商朝 Shāngcháo or Shang Dynasty (16th century BC–c. 1046 BC) the Chinese were accomplished masters at bronze casting. There are many examples of 商朝 Shāngcháo bronze objects in museum collections all over the world and they have become a symbol or icon of that age. A recreation of a bronze 爵 jué is shown below in Figure 3.14.

Figure 3.14: A recreation of a 商朝 Shāngcháo bronze 爵 jué.

The 爵 jué is a ritual object used to serve warmed wine. For readers interested in bronze, the British museum website has an extensive collection of bronze vessels. Images of a 爵 jué can be found at the following link: https://research.britishmuseum.org/research/collection_online/collection_object_details.aspx?objectId=259705&partId=1&searchText=bronze+jue&page=1.

What is so special about bronze? We previously mentioned that an alloy can have properties that are different from the constituent elements. Bronze is an alloy of primarily copper 铜 tóng and tin 锡 xī. Both copper and tin are soft metals. Tin melts at only 232°C while copper has a much higher melting point of about 1085°C. When mixed together they produce a new metal that is very hard. Bronze alloys can be formed with a very wide range of copper to tin ratios. There are ancient examples of bronze with copper content of only about 50% all the way up to about 90%.

The melting point of bronze is generally lower than that of copper, but the exact value depends on the elemental composition. Molten bronze tends to flow much better than pure copper and that makes it advantageous for casting intricate patterns. In our discussion about the properties of metals we pointed out that metals are malleable and ductile but bronze is not. Bronze is something we call a substitutional alloy. This type of alloy is formed by replacing some atoms of the parent material with a small percentage of atoms of a different element. Copper atoms can be arranged in a structure called a crystal. If we take a particular orderly arrangement of atoms and repeat it over and over again we form a crystal. Symmetry is an important requirement for the atomic arrangement. In nature we find that there are six basic crystal families. These families are known as triclinic, monoclinic, orthorhombic, tetragonal, hexagonal, and cubic. Cubic crystals have the simplest arrangement and many minerals take on this form. In a simple cubic crystal, each atom is arranged at the corner of a cube. We can make the arrangement slightly more complex by placing another atom in the center of the cube. This form is known as body-centered cubic (BCC). If we start with a simple cubic (SC) structure and add atoms on each of the crystal faces we get a face-centered cubic (FCC) structure. Copper forms an FCC structure. Various forms of cubic structures are shown in Figure 3.15.

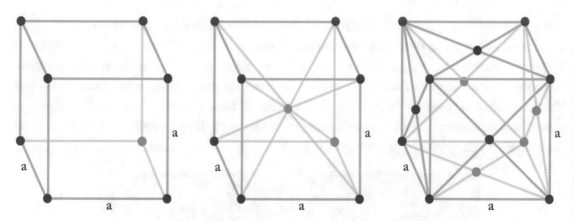

Figure 3.15: Cubic crystal structures. Simple cubic, body-centered cubic, and face-centered cubic.

To form a substitutional alloy such a bronze we replace some fraction of the atoms with another atomic species. This is shown below in Figure 3.16. Tin is soluble in copper and we can add up to about 13% by weight while still preserving the face-centered cubic structure. The addition of tin does change how easily the crystal planes can slip past each other in response to an external force. Adding tin makes the bronze harder by restricting the ease of slip. Too much tin makes the material brittle. A bronze edged weapon is much harder than one made of pure copper. However, if the tin content is too high, the blade will become brittle and is easily broken.

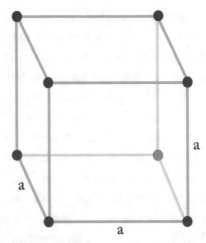

Figure 3.16: A substitutional alloy in which an atom has been replaced with a different atomic species of about the same size (black dot).

3.8.1 BRONZE COINS

China's hunger for copper rose out of its need to create bronze. Bronze for weapons, works of art, ceremonial objects, and their enormous monetary system. Figure 3.17 shows some examples of ancient Chinese bronze money from the author's own collection. China has been casting bronze coins since about 600 BC. Early coins had a value based on their weight such as the 半 两 bàn liǎng or half a liang coin shown in Figure 3.17 (b). The 两 liǎng is a unit of mass and is sometimes called the Chinese ounce. At the time the coin was cast, the 两 liǎng was about 16 g compared to the modern 两 liǎng which has a mass of 50 g.

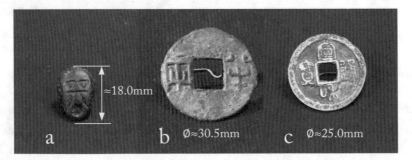

Figure 3.17: Three forms of ancient Chinese money made of bronze. (a) 蚁鼻钱 yǐ bí qián sometimes called "ant nose money," Warring States period (400–220 BC). (b) 半 两 bàn liǎng, Qin State (ca. 300 BC). (c) 宣咊通宝 xuānhé tōng bǎo, Northern Song Dynasty (960–1127).

Common people may not have had the means to own a bronze ritual object but they often carried around some of the emperor's "treasure" in their pockets and purses. During the reign of the Emperor 唐高祖 Táng Gāozǔ, founder of the 唐朝 Tángcháo or Tang Dynasty, a new type of bronze coin was introduced. Prior to the new coin's introduction in 621, coins often had an inscription indicating their value based on weight. This new coin bore the inscription 通宝 tōng bǎo which can be translated as "circulating treasure," as shown in Figure 3.18.

Figure 3.18: Bronze 开元通宝 Kāiyuán tōngbǎo coin of the Tang dynasty. Introduced in the year 621.

Bronze used for coins was not just a mixture of copper (Cu) and tin (Sn). Lead (Pb) was often added to the mix. This tends to improve the way the molten bronze would flow when poured. For example, the coin shown below in Figure 3.19 is similar to a specimen in the British Museum (Item number 1883,0802.508).

Figure 3.19: Bronze 皇宋 Huángsòng tōngbǎo coin of the Northern Song dynasty.

The composition of the 皇宋 Huángsòng coin in the British Museum collection is Cu 71%, Sn 12%, and Pb 16% (Wang, Cowell, Cribb, and Bowman, 2005). Small amounts of other elements are also present but these are the main constituents. 沈括 Shěn Kuò would have recognized this coin, since it is from the 北宋朝 Běi Sòng cháo or Northern Song dynasty and was issued in the year 1039, about 8 years after he was born. It has been estimated that between the years 960 and 1125 the 北宋 Běi Sòng government minted approximately 200,000,000 strings of cash. A string of cash nominally consisted of some 900–1,000 coins. So just how much copper did China need to mint those coins? In physics we often perform order of magnitude calculations. It is often impossible or very difficult to find exact values for all the terms in an equation but we can arrive at an intelligent estimate of the result within the nearest power of ten. Powers of ten represent orders of magnitude. The ability to make such calculations is important when we try to estimate a physical quantity.

We know that the mass of coins varied a great deal but for the purpose of this order of magnitude calculation we will assume that the coin in the museum represents the average for that period. The mass of this coin is 3.27 g, which we will round off to 3.3 g. From the British Museum data it is clear that most coins from this time period are between 3–4 g so we will just use 3.3 g for the calculation. The number of coins in a string has varied at different times but we will assume that one string is 1.0×10^3 coins. So, for 200,000,000 or 2.0×10^8 strings of coins we have,

$$\frac{1.0 \times 10^3 \, coins}{string} \times 2.0 \times 10^8 \, string = 2.0 \times 10^{11} \text{ coins.}$$

Now we can calculate the mass of copper required. We will use a copper ratio of 70% as an estimate. So, a single coin of mass 3.3 g contains

$$0.70 \times 3.3 \text{ g} = 2.3 \text{ g.}$$

For the estimated 2.0×10^{11} coins, the mass would be

$$m = 2.0 \times 10^{11} \, coin \quad x \quad \frac{2.3 \, g}{coin} = 4.6 \times 10^{11} \, g \,.$$

This is a huge number and it would be more convenient to express it in metric tons using the conversion $1t = 1.0 \times 10^3$ kg.

$$m = 4.6 \times 10^{11} \, g \times \frac{1.0 kg}{1.0 \times 10^3 \, g} \times \frac{1.0 t}{1.0 \times 10^3 \, kg} = 4.6 \times 10^5 \, t \,.$$

That is almost half a million tons of copper! What would that be worth in today's money? In early December of 2019 copper sold for about $2.73/lb (pound weight).

$$price = 4.6 \times 10^{11} \, g \times \frac{1.0 kg}{1.0 \times 10^3 \, g} \times \frac{2.2046 \, Lb}{1.0 \, kg} \times \frac{\$2.73}{Lb} = \$2.8 \times 10^9.$$

That is 2.8 billion US Dollars!

This output does represent a period of 165 years, but it is still an impressive accomplishment for nearly 1,000 years ago. When dealing with huge numbers it is often helpful put them in some historical context and make comparisons. During the 北宋朝 Běi Sòng cháo annual copper production peaked at a rate of 1.3×10^4 t/yr (Jost, 2014). From 1801–1810, the entire worldwide production of copper was 9.1×10^4 t for an average annual rate of 9.1×10^3 t, which is less than the 北宋 Běi Sòng peak , some 850 years prior. In 1900, the entire world production of copper was 4.96×10^5 t, slightly more than the entire 北宋朝 Běi Sòng cháo period ("Copper," 1910). Of course the introduction of electric power and electronic devices drove the need for copper in the modern age. China is now one of world's largest producers of copper. As of 2017, China's annual production of 1.656×10^6 t was the third largest in the world. The top producer in 2017 was Chile with an annual production of 5.503×10^6 t (World Mining Congress, 2019, p. 116).

3.9   WET COPPER METHOD

China's production of copper in the 北宋朝 Běi Sòng cháo was not sustainable. Near the end of the 11th century the government was looking for other sources of copper. In the 1090's a government official by the name of 游經 Yóu Jīng took a special interest in the 胆水浸法 dǎnshuǐjìnfǎ or vitriol water steeping method as a way of supplementing the declining output. Throughout this text, we shall call this process "wet copper." There is very little data available about the amount of wet copper produced. Alexander Jost, who has researched this topic extensively, points out that between the years 1111–1118 the wet copper process provided about 27% of China's annual production quota. That translates to a yearly production of 4.66×10^3 t. This was short lived, and by 1162 overall copper production dropped to a mere 1.74×10^2 t, but wet copper provided most of that contributing 1.41×10^2 t to the total (Jost, 2014). As wet copper production was tapering off and resources were being depleted China was to face a far more serious problem than copper scarcity. The early 13th century brought the Mongol invasion and their insistence on using paper money. Bronze currency did not completely disappear from China but the need was drastically reduced. The history of Chinese currency after Mongol rule is complex and we will not go into the details. Coins made from mostly copper, bronze, and brass were used up until the time of the founding of the republic of China in 1912. Brass is a copper alloy that uses zinc instead of tin or is occasionally mixed with tin. It is also a substitutional alloy and has good corrosion resistance.

As Figure 3.20 attests the wet copper method is still practiced in China today, but it is miniscule compared to the giant open pit mines currently used.

Figure 3.20: A wet copper mine showing flumes filled with scrap metal onto which the copper precipitates. The bucket contains copper "mud" which is rinsed from the metal surface. Photo courtesy of Alexander Jost.

As you can see from Figure 3.20, the wet copper process has some distinct advantages. This method does not require digging complex and dangerous mine shafts which saves on human labor. There is no need for thermal processing until the last step when the copper mud is melted to form an ingot. This provides a huge savings in firewood or other fuels. Wet copper also allows the exploitation of copper that would otherwise remain inaccessible to human tunneling. Water can be diverted into copper bearing earth that is difficult to access or that has too low of a yield for more conventional mining methods. Setting up a wet copper production facility requires very little capital equipment cost. Of course, one has to have access to a water source with enough copper to make it worthwhile. Settling ponds and flumes are very easy to construct compared to the cost and danger of underground tunnels. Every few days, depending on the copper concentration, workers rinse the precipitated copper from the metal surfaces. Copper is easily rinsed off the scrap metal and forms a "mud" which is collected in buckets, as shown in Figure 3.20. Mining is a complex art. One can easily spend more money smelting the copper to make coins than the actual value of the coins themselves. This situation actually did happen from time to time in ancient China.

沈括 Shěn Kuò was not the first to report the wet copper process nor did he implement it as an industrial scale production method. 张潜 Zhāng Qián (1025–1105) is often credited as the discoverer or father of the wet copper process. Figure 3.21 shows a statue of 张潜 Zhāng Qián ladle in hand ready to harvest copper. There is a reference dating to the 2nd century BC that sounds like a similar reaction but it was not used on a large scale to produce copper.

Figure 3.21: 张潜 Zhāng Qián (1025–1105) credited as the discoverer the wet copper process. Photo courtesy of Alexander Jost.

In his book *A Summary of Copper Steeping* 浸銅要略 Jìntóng yàolüè, 张潜 Zhāng Qián describes the essential elements of the wet copper method. This information was presented to the Emperor 哲宗 Zhé Zōng who reigned from 1085–1100 by 张潜 Zhāng Qián's son named 張甲 Zhāng Jiǎ. 沈括 Shěn Kuò served the Emperor 神宗 Shén Zōng, who was the father of 哲宗 Zhé Zōng. He also served for a short time under 哲宗 Zhé Zōng but was falsely charged. This was actually the second false charge brought against him, the first came under the previous emperor. In the fifth year of the reign of 哲宗 Zhé Zōng, 沈括 Shěn Kuò left the capital city went into what some might call retirement. It was during that period of retirement that he began to write down his experiences in his great work 梦溪笔谈 Mèng Xī Bǐtán, known in the West as *Brush Talks from Dream Brook*. As we have mentioned in Chapter 1, 梦溪笔谈 Mèng Xī Bǐtán is considered a treasure trove of early scientific reports. Some of the items were retelling of information he picked up over the years and some were based in personal experience. In the year 1095, 沈括 Shěn Kuò died of an illness.

3.9.1   CHEMISTRY OF THE WET COPPER PROCESS

In this section we will explore the chemistry behind the wet copper process. We have already addressed 沈括 Shěn Kuò's explanation based on the five phases or 五行 wǔxíng theory. In that section we mentioned that the "bitter water" was likely some type of copper salt. In chemistry the term salt does not just mean table salt but rather a compound that contains positively charged ions called cations and negatively charged ions known as anions. Table salt or sodium chloride (NaCl) is composed of the cation sodium (Na^+) and the chloride (Cl^-). Copper sulfate has the copper cation Cu^{2+} and the sulfate anion SO_4^{2-} which forms the compound $CuSO_4 (H_2O)_x$ where x can range from 0–5. Anhydrous copper sulfate corresponds to x = 0 and the crystals are gray/white. Copper sulfate pentahydrate corresponds to x = 5 and produces the beautiful blue crystals as shown in Figure 3.10. When small amounts of copper sulfate pentahydrate $CuSO_4 \cdot 5H_2O$ are dissolved in water the strength of the blue color varies with concentration. This is shown in Figure 3.22.

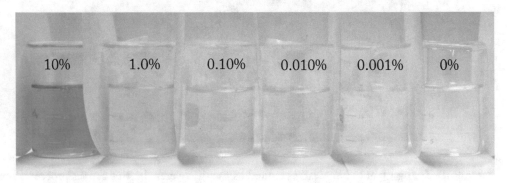

Figure 3.22: Copper (II) Sulfate Pentahydrate $CuSO_4 \cdot 5H_2O$ crystals dissolved in distilled water to illustrate color differences. Concentration given in w/v (%). The last beaker (0%) is only water.

In Ancient China, prospectors used taste, smell, and color to identify water sources for wet copper production. It is hard to imagine that they could discern concentrations below 1% w/v by color especially in the presence of rocks and organic matter found in streams. In Chapter 5 there is an activity in which we treat copper sulfate pentahydrte as an ore and mine copper from it. We can calculate the amount of copper in solutions of 10% w/v and 1% $CuSO_4 \cdot 5H_2O$ and then compare those results to the amount of recoverable copper in geological ore sources. As you can see from the graph in Figure 3.23, the amount of copper recoverable in ore has been decreasing as the richer ore has been depleted. The 1–10% w/v solution window brackets the copper concentration in geological ore.

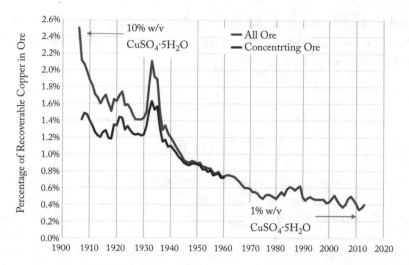

Figure 3.23: Average grade of U.S. Copper ore with copper content of Copper (II) Sulfate Pentahydrate solutions overlaid for comparison. Plazak (https://commons.wikimedia.org/wiki/File:Grade_of_US_Copper_Ore.png), overlay data by Matt Marone.

3.9.2 WET COPPER DISPLACEMENT REACTION

As a physicist or chemist we can explain the wet copper process as an example of a displacement reaction. In a displacement reaction one element replaces another element in a compound. For the wet copper process we can write the chemical reaction as:

$$Fe(s) + CuSO_4(aq) \longrightarrow Cu(s) + FeSO_4(aq).$$

Solid iron (Fe) is immersed in copper sulfate ($CuSO_4$). Notice that in the product we now have soild copper formed and the iron has switched places forming iron sulfate. This is also an example of an oxidation-reduction reaction. In an oxidation-reduction reaction, also called a redox reaction, the oxidation state of the atoms involved is changed. Once again, it is the dance of the electrons at work. Electrons are transferred between the atoms. This transfer of electrons can be described by the mnemonic "Oil Rig." This is a mnemonic that reminds us:

Oxidation **is l**oss of electrons and **R**eduction **is g**ain.

In order for something to be reduced there must be a corresponding oxidation. The ionic equation describing the wet copper process is given by:

$$Fe(s) + Cu^{+2}(aq) \longrightarrow Fe^{+2}(aq) + Cu(s).$$

Cu^{2+} is reduced by gaining two electrons and will plate out at the site where the reduction takes place, that is the iron metal. The iron metal started out as a neutral solid or Fe^0. Now it has become Fe^{+2} by losing two electrons, so it is oxidized. That is,

$$Cu^{2+} + 2e = Cu^0$$

$$Fe^0 - 2e = Fe^{2+}.$$

Remember, the sulfate anion is SO_4^{2-}, so the iron combines with the sulfate to form $FeSO_4$.

3.10   ELECTRONEGATIVITY

Will any metal that is immersed into the liquid be plated over with copper? Why not use some other metal for the process? Figure 3.24 shows the results of an experiment designed to answer that very question. In the upper photo we show several different metals arranged on a glass well plate. Then, 2–3 mL of Copper (II) Sulfate Pentahydrate $CuSO_4 \cdot 5H_2O$ is added to each well. The lower image shows what happens after about 5 minutes of exposure to the solution. Most of the solution is pipetted away for the photograph. The metals in the top row: Gold (Au), Platinum (Pt), and Silver (Ag) are left unchanged. The middle row is copper which serves as a reference. The bottom row of metals consists of Iron (Fe), Indium (In), and Zinc (Zn). These metals have reacted but with different degrees of activity. The Zinc sample stands out as having the most metal deposited on it. Next to each metal is a number that is a measure of the elements electronegativity (χ). Notice that the metals with electronegativity greater than that of copper are unchanged. Metals with an electronegativity lower than that of copper are coated. The electronegativity values shown are measured on the Pauling scale and are unitless quantities. Zinc has the lowest electronegativity at $\chi = 1.65$ compared to copper at $\chi = 1.90$. Gold is an unreactive metal and, in this experiment, has the highest electronegativity at $\chi = 2.54$ of the metals shown. The higher the electronegativity, the more an atom attracts electrons toward itself. Copper is only reduced when the other metal is oxidized or loses two electrons. This happens much more easily with iron, indium, and zinc because their electronegativity is lower than that of copper.

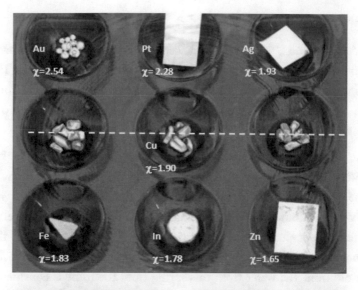

Figure 3.24: Metals of differing electronegativity before (top) and after (bottom) 5-minutes exposure to 10% w/v Copper (II) sulfate pentahydrate solution. Electronegativity values are given on the Pauling scale.

Zinc was not used in ancient China although it is found associated with other metals such as iron in a sulfide ore known as Sphalerite. Due to zinc's low melting point, it was not isolated as an element until the year 1746. Zinc does show up in Chinese coins but not until the early 1500's during the 明朝 Míng cháo or Ming Dynasty. The tin content of these coins is significantly reduced and zinc content increases to form brass. Indium can also be found in Sphalerite but was not discovered until 1863.

3.11   WET COPPER MINING IN ARIZONA

What do Jerome Arizona and 沈括 Shěn Kuò have in common? Jerome Arizona was home to the United Verde Mine which produced gold silver and copper. Mining ceased in 1953 but this area was a major mining center in the 19th century. Below is an excerpt from the journal *Gleanings in Bee Culture* (Root, 1897). This description from 1897, lays out the wet copper process in the folksy language of the day:

> Well, one of the most wonderful things about the Jerome gold-mine is a spring of water that runs out in considerable quantity from the lower drift. The water runs out beside the track. It is carried into a wooden flume something like the irrigating-flumes; and this wooden flume runs along the mountain side pretty nearly level for a mile or more. The flume is perhaps ten or twelve feet wide, and the water in the bottom is several inches deep. Now, on the bottom of this flume they have laid all sorts of pieces of refuse old iron. The water from this spring from the mine is considerably impregnated with sulphate of copper, or blue vitriol, as It is generally called. Perhaps many of our friends have observed that, when they dip a bright piece of iron or steel-say a knife blade- into a solution of sulphate of copper, the blade soon becomes coated or plated with copper. When you are spraying fruit-trees with the copper sulphate, you may have noticed this. The explanation is that the sulphuric acid has a stronger liking for the iron than for the copper; so it lets go of the copper, as it were, and grasps hold of the iron, which is an easy solvent. The copper must go somewhere, so it is left on the surface of the iron. Cast-iron articles are often copper-coated by this means. Well, at this Jerome mine the copper is held in solution in such quantities that the iron causes it to drop the copper, not only all over the iron articles, but even on the bottom of the wooden flume. Every little while this loose mass of copper dust or mud is shoveled up. When they get a carload or more it is melted down, and it gives ingots of almost pure copper. Strangely enough-at least it was strange to me-this precipitated copper also contains a percentage of gold; and my friends told me that a sharp Yankee down by the Verde River had commenced speculating on his own hook by precipitating an additional quantity of copper from Copper Creek after the Jerome mine had got through with it and let it go to waste. Before the mine was ever discovered, people knew this spring and called it Bitter Spring because no man or animal could drink the water. Now this spring yields a mint of money when you get it out, by chunks of old iron, in the way I have described.

Was this technology independently discovered in the American west or perhaps imported from the Chinese? In the 19th century, many Chinese known as "coolies," performed hard labor in the American west. They lived in brutal slave-like conditions, and unfortunately their history has been largely ignored. The mining and railroad industries employed a large pool of such laborers. Could they have brought the knowledge of wet copper mining with them? We do not know for sure. As for their harsh labor, the term "coolie" sounds like it might be a loan word from the Chinese term 苦力 kǔlì, which means bitter labor.

REFERENCES

An, L. (2010). *The Huainanzi*. (J. S. Major, S. Q. Meyer, and H. D. Roth, Trans.) New York: Columbia University Press. 96

Cheng, T.-k. (1974). Metallurgy in Shang China. *T'oung Pao*, 60, 209–229. DOI: 10.1163/156853274X00063. 89

Copper. (1910). In *Encyclopædia Britannica* (Vol. 7, p. 109). New York: The Encyclopædia Britannica Company. 105

Golas, P. J. (1999). *Science and Civilisation in China, Chemistry and Chemical Technology* (Vol. XIII). Cambridge University Press. 94

Jost, A.F., (2014). From Secret Knowledge to Mass Production: The Wet Copper Industry of Song China 960–1276 [Unpublished doctoral dissertation]. Eberhard-Karls-Universität Tübingen. 105

Mei Jianjun (2000). *Copper and Bronze Metallurgy; in Late Prehistoric Xinjiang: Its Cultural Context and Relationship with Neighbouring Regions*. Oxford: Archaeopress. DOI: 10.30861/9781841710686. 85

Root, A. (1897, April 15). Notes of Travel. *Gleanings in Bee Culture*, pp. 297–300. DOI: 10.1038/056297a0. 112

Wang, D. Q. and Carey, M. C. (2014). Therapeutic uses of animal biles in traditional Chinese medicine: an ethnopharmacological, biophysical chemical and medicinal review. *World Journal of Gastroenterology*, 20(29), 9952–9975. DOI: 10.3748/wjg.v20.i29.9952. 94

Wang, H., Cowell, M., Cribb, J., and Bowman, S. (2005). *Metallurgical Analysis of Chinese Coins at the British Museum*. Publication No. 152. London: British Museum Research. 104

World Mining Congress. (2019). *World Mining Data* (Vol. 34). Retrieved December 10, 2019, from http://www.wmc.org.pl/sites/default/files/WMD%202019%20web.pdf. 105

Experimental Section

DESCRIPTION OF EXPERIMENTS

A distinctive characteristic of the class we teach at Mercer University is the hands-on laboratory component. Not only do our students learn about the history of Ancient China, they also recreate several notable technologies. In the previous book about this topic, *Teaching Physics through Ancient Chinese and Technology*, we provided several experiments related to force, torque and statistical treatment of data. In this volume we have included an experiment on magnetic fields and two methods of "mining" copper.

The first experiment is making a magnetic compass and is described in Chapter 4. Our procedure is based on a description given by 沈括 Shěn Kuò in his famous work 梦溪笔谈 Mèng Xī Bǐtán also known as *Brush Talks from Dream Brook*. Not only do we make a magnetic compass but we also use it as a simple magnetometer. We use the compass/magnetometer to measure the strength of the magnetic field produced by a long straight wire. The magnetic field should be proportional to the current through the wire and inversely proportional to the distance between the compass and the wire. Students investigate there proportionalities and learn techniques of data analysis.

In Chapter 5, we look at two methods of mining copper. As we have pointed out in Chapter 5, copper is used to make bronze, the metal that changed the ancient world. First, we experiment with reducing the mineral malachite to elemental copper. Malachite beads can be found in many hobby supply shops and are an inexpensive source of copper ore. Our second experiment illustrates the ancient technique of wet copper mining. In this experiment we mine copper from commonly available copper sulfate pentahydrate $CuSO_4 \cdot 5H_2O$. This experiment introduces students to some simple chemical principles such as displacement reactions, oxidation, reduction and electronegativity. The wet copper experiment does not uses open flames or high temperatures and is inherently safer than the malachite smelting technique.

CHAPTER 4

Compass 指南针 Zhǐnánzhēn

4.1 OBJECTIVE

The objective of this experiment is to study the magnetic properties of various materials using a homemade compass. We will also use the compass to study the magnetic field produced by currents through wires.

Chinese Vocabulary		
Lodestone or magnetite 磁石 císhí		
Magnet 磁铁 cítiě	Iron 铁tiě	
North 北 běi	South 南 nán	
Compass 指南针 zhǐnánzhēn (pointing south needle)		
Compass used for fēngshuǐ (风水), 罗盘 luópán		
Materials		
Compass Making:		
Lodestone	Wax (soft at room temperature)	Glue stick
Pins	Protractor image sized to Petri dish	White marker or paint
Petri dish	Alcohol lamp or lighter	Scissors
Diagonal cutter	Steel wire, such as music wire	Safety glasses
Silk or fine thread	Strong magnet	
For magnetic field measurements:		
Power supply > 3A	Ammeter	Beaker to shield power resistor
Alligator clips	Non-conducting support for rod	
Hook up wire	2Ω, 25W power resistor	
Non-magnetic conducting rod about .5m length		

4.2 INTRODUCTION

The seemingly strange properties of the lodestone have been known since ancient times and many myths exist telling about its seemingly supernatural powers. Lodestones and magnets seem magical in that they exert a force on an object without making actual physical contact. This is very different from the physical principles we have studied thus far. When we studied the acceleration of an object there was a force acting on the object due to a push or a tension in a rope. Some sort of physical contact existed between the object that was being accelerated and the object applying the force. When we discussed weight we said that it was due to the gravitational force acting on the object. This explanation takes the existence of the gravitation field for granted. Now we will study the magnetic field and how it exerts a force on an object without a physical contact. A more detailed discussion of magnetism can be found in Chapter 1. As we have already mentioned in Chapter 1, one of the earliest known references to magnetic compasses makes reference to magicians or adepts who use the magnetic needle. This reference comes to us from the brush of the famous Song dynasty (960–1279) polymath 沈括 Shěn Kuò. Shěn Kuò was an expert in many different fields of science history and politics and it rightfully considered the greatest scientist in the long history of China. Some 400 years before William Gilbert explained to the Europeans that the Earth was a giant magnet whose magnetic and geographic poles were not aligned, 沈括 Shěn Kuò wrote about magnetism and lodestones in his memoirs. These memoirs, written in about 1088, known as 梦溪笔谈 Mèng Xī Bǐtán or *Brush Talks from Dream Brook* contain a wealth of information about science, history, culture, and politics.

> Adepts can make a needle point to the south by rubbing it with a magnetic stone. However, the needle often slants to the southeast, not pointing due south. If the needle is floating on water it is unsteady. The needle may be placed on a fingernail or the edge of a bowl, which will make it turn more easily, but since these supports are hard and smooth, it may easily fall off. The best way, is to suspend the needle by a single fiber of raw silk from a cocoon, which is attached to the center of the needle by a small piece of wax the size of a mustard seed. Then, hang it in a windless place and it will often point south. Needles may sometimes point North after being rubbed. In my house, I have needles that point North and others that point south. The south pointing property is like the habit of cypress trees that always lean to the west. Nobody knows the reason why it is so.

Shěn Kuò tells us that a needle is rubbed on a lodestone. 磁石 císhí or lodestone is a naturally occurring magnet that is made from the mineral magnetite. We will begin or experiment with lodestones like the one shown in Figure 4.1.

Magnetite is an oxide of iron and has been know from very ancient times in a wide range of cultures. 沈括 Shěn Kuò does not mention it, but a lodestone itself can be suspended from a string and used as a compass. Since it is an irregularly shaped rock there is no sharp point to act as an indicator.

Figure 4.1: 磁石 císhí or lodestone attracting a steel needle.

Figure 4.2: A lodestone orienting itself in the magnetic field of the Earth acting like a compass needle. Note the "S" indicating the south pointing side.

Returning to the text, notice we are told that the compass needle points south. In European culture we normally think of a compass needle as pointing north. South is an important direction in Chinese culture. The Emperor sat with his back to the north and faced south towards the sun. Anyone approaching the emperor did so from the south. Honored guests were seated so that they faced south during a meal. Houses were often built with a southern exposure to gain warmth provided by the sun. We should also bear in mind that the earliest use of the magnetic compass was not for navigation but for fēngshuǐ 风水. Determining the auspicious direction to build a house, or a city was the job of the Geomancer aided by his special compass used for 风水 fēngshuǐ, known as a 罗盘 luópán.

The next important fact we encounter is that the needle inclines slightly to the east as illustrated in Figure 4.3. That is slightly to the east for a south pointing needle, or slightly to the west for a north-pointing needle, as shown below. This deviation from true north is known as magnetic declination and is discussed in more detail in Chapter 1.

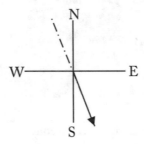

Figure 4.3: Magnetic declination of a compass needle.

This observation that the compass needle does not point directly south is one of the earliest references to what we now call magnetic declination. The current magnetic declination of Macon, Georgia is about 5° W. This means that a north-pointing needle would not point due north, but rather, 5° west of north. The map shown below indicates the position of the North Magnetic Pole and how it wanders with time.

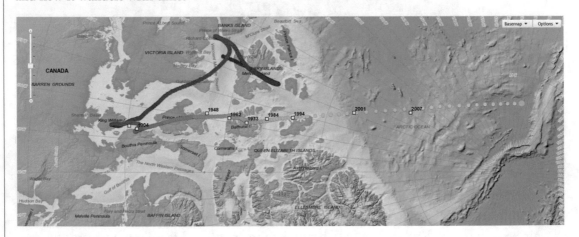

Figure 4.4: Wandering of the North Magnetic Pole. Image from the National Centers for Environmental Information (NCEI). Yellow squares are observed data; circles are modeled results dating back to 1590 (blue).

4.3   LODESTONE EXPERIMENT

For our first experiment we will try to find the magnetic poles of our lodestone. Use some thread and hang the lodestone so that it can rotate freely. Once the stone has oriented itself, mark the poles with whiteout as shown below.

Figure 4.5: Allow the lodestone to orient and mark the south-pointing end.

Now bring some objects near the lodestone and see if they have any effect on its orientation. Record your results in the data table and comment on this in your lab report.

4.4   COMPASS CONSTRUCTION

Now we will make a simple magnetic compass similar to the one 沈括 Shĕn Kuò described. We will use steel music wire for the needle, wax, and silk for the support and a Petri dish to provide a "windless place." The large diameter part of the Petri dish will be the bottom of the compass which will have a protractor scale for measuring the orientation of the needle. A completed compass is shown below.

Note that the south-pointing end of the wire is marked with white paint. The length of the steel wire is trimmed with diagonal cutters. This wire must be long enough to be close to a scale marking but not too long or it will strike the wall of the dish. This steel wire will serve as our compass needle. Stroke the steel wire along the long axis with a lodestone several times to magnetize it and turn it into a compass needle.

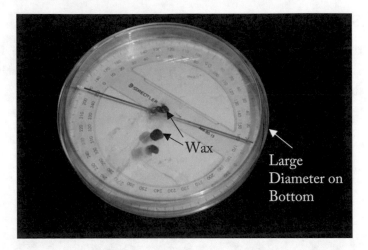

Figure 4.6: Completed compass.

We will now make a hole in the center of the top plate. The silk thread that supports the needle will pass through this hole. We must position the hole as close to the center as possible so that it will be aligned with the protractor scale on the bottom plate. Figure 4.7 shows an easy way to find the center. Simply balance the cover on the point of a pin as shown below.

Figure 4.7: Find the center by balancing the lid on a pin.

Once the plate is balanced, press the pin to make an indentation into the plastic. **Do not** try to push the pin through the plastic or else **it will shatter**. The best way to make the hole is to melt the plastic with the end of a hot piece of wire or pin. You can use an alcohol lamp or a lighter to heat the wire and press it through the plastic at the center point mark. Do not try to drill the plastic as it will cause the lid to shatter.

A close-up image of the top plate is shown below. Place a small bit of wax onto the compass needle and squeeze it to make an attachment point for the silk. Since the wax is sticky this will enable us to connect the steel wire to a fine silk thread. Note that the silk thread **passes through** the small hole in the top plate. **Do not** fill the hole with wax. Several more bits of wax are used to attach the free end of the silk thread to the top plate. Leave some of the silk thread extending past the wax. This will help when it is time to adjust the height of the needle.

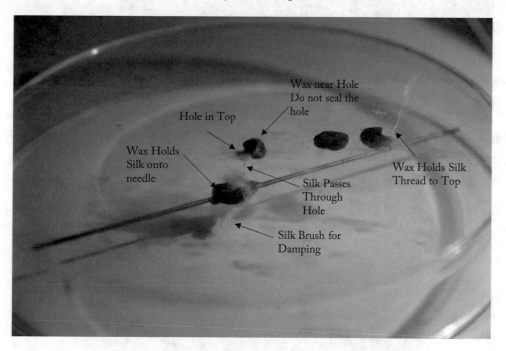

Figure 4.8: Close-up of top cover showing position of wax and silk thread.

Now we will attach a small thread to the wax on the needle. This part is a bit tricky. Pluck a small amount of silk and try to pull out a few long fibers. You just need a few fibers to make a very fine thread. A thick thread is not good because it will take more force to turn the wire.

Once you have a short length of silk thread, you will pass it through the hole and attach it to the wax joint on the needle. A few more short fibers can be attached to form the damping brush. Once the needle is suspended it will swing like a pendulum and take a long time to stop moving. In order to make the swinging die down quickly add a few fibers to the wax and form a brush. This brush should just gently contact the bottom Petri dish otherwise it will interfere with the motion of the needle. The wax suspension joint is shown below. When handling these fibers it is best to use a dark background such as the stone top lab tables. It will be extremely difficult to see the fibers against a light-colored background.

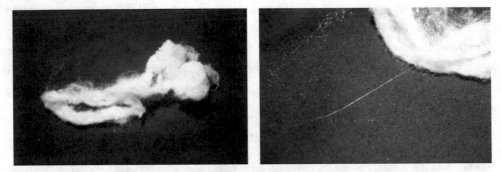

Figure 4.9: Pull out a few fibers of silk.

Notice that only a few fibers pass through the hole and the damping brush hangs below the needle. After you pull the silk through the hole secure it to the top plate with a small bit of wax, but do not fill the center hole. Now we get to the most tedious part of the procedure.

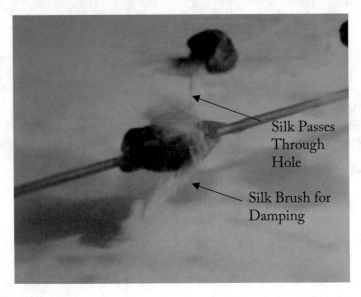

Figure 4.10: Adjust the balance of the needle by slightly moving it through the wax joint.

The needle should be balanced level and centered so that it does not scrape against any part of the Petri dish. Place the top plate at the edge of the desk so that the needle can move freely. Slide the needle in the wax joint so that is hangs level, as shown in Figure 4.10. This will take a few attempts, don't get frustrated. Remember you are balancing the weight of the needle and the magnetic dip in you location. Once the needle is level be sure that it orients itself in the magnetic field and points like a compass. If it does not orient properly you may need to use a more powerful magnet to magnetize the needle.

Once you are satisfied that the needle is level and orients in the magnetic field, you can put the top plate onto the bottom and close the dish. Your needle may be sitting on the bottom of the dish. Remember the wax bits that hold the silk thread in place on the top plate? You can now adjust their position to pull the needle up so that it is not sitting on the bottom of the dish. You must carefully adjust the height so that the damping brush makes light contact with the bottom plate. If it is not in contact, the needle will take a long time to settle down. If the brush just touches the bottom plate, the needle motion will stabilize more quickly.

Now it is time to attach the protractor scale to the bottom plate. The scale is cut out with scissors glued to the bottom of the dish on the **outside surface.**

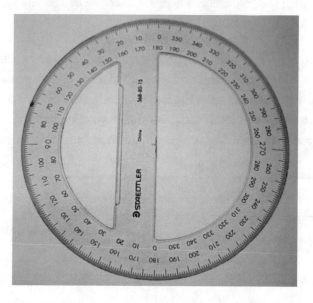

Figure 4.11: Protractor scale to be mounted in the bottom of the dish.

Alignment is critical. We want the center of the scale to line up with the center hole as accurately as possible. Glue the scale to the **outside of the dish** and pay special note to the alignment marks as shown below.

If the scale is aligned as indicated, the center of the scale should be properly positioned. The bottom part of the Petri dish can be rotated to align the scale with the needle. This will be important when we use the compass to make measurements. Align the scale with two marks 60° from either side of zero and one at 180°.

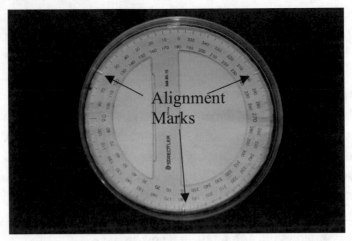

Figure 4.12: Align the scale with two marks 60o from either side of zero and one at 180o.

Once the scalc is glued in place on the outside of the bottom plate, assemble the compass in its final form. Check to make sure your compass reacts freely. Place the lodestone near the compass and check to make sure it reacts to the magnetic field. Find a few objects made of different materials and bring them up close to the compass or the lodestone. Do they cause it to deflect? Record your observations in the data sheet and comment about this in your lab report.

If all goes well you have just made a magnetic compass! Amaze your friends; use it to navigate around campus.

Data Sheet for Lodestone response	
Group members_____	
Object	**Compass/Lodestone Response**

4.5   MAGNETIC FIELD PRODUCED BY A LONG STRAIGHT WIRE

We can now use our compass as a laboratory instrument to measure the magnetic field B of a wire. Prior to the time of Ørsted's discovery of electromagnetism (published April 21, 1820) magnetic fields were only known to be produced by naturally occurring materials. Ørsted discovered that a current passing through a wire also produces a magnetic field. This "artificial" means of producing a magnetic field has led to a great number of useful technologies such as electric motors and MRI. As we have discussed in Chapter 1, the magnetic field produced by a long straight wire is in the form of concentric circles with a direction given by the right hand rule. The strength of the magnetic field is given by the equation

$$B = \frac{\mu_0 I}{2\pi R},$$

Where I is the current through the wire and R is the radial distance from the wire to a point in space where the magnetic field acts. In this experiment we will explore the proportionalities described by this equation. We will first explore how the strength of the magnetic field is related to the current through the wire. Current is passed through the rod by connecting the circuit shown in Figure 4.13. Note the 2 W power resistor can get very hot so avoid touching it and place it on a non-flammable surface. We often place the resistor in a beaker for safety so that students will not accidently touch the hot surface.

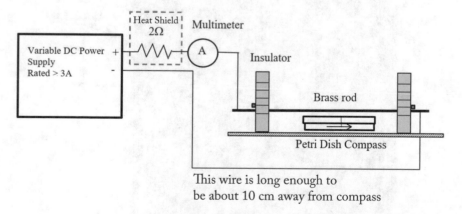

Figure 4.13: Circuit for magnetic field experiment.

4.6 THE COMPASS AS A MAGNETOMETER

In order to measure the strength of the field, we must do a little geometry. The compass needle aligns itself along the resultant magnetic field. In our setup, the resultant field has two components. One component is the magnetic field of the Earth and the other component is the magnetic field produced by the wire. We start by aligning the wire so that it runs in the same direction of the Earth's field as indicated by the compass needle. The wire (rod) should be placed directly above the compass needle with no current through the wire. In order to align them properly, **you must look straight down** onto the rod and move it so that it lines up with the needle. Also rotate the compass scale so that the needle lines up with 0 at one end and 180° at the other end. Whenever you make an angle measurement, read the angle on both sides of the compass scale so as to minimize parallax error. If the needle deflects counter clockwise by 10° on the south end, then the north end should also deflect 10°. If the readings is not symmetric change your viewing angle.

When current flows through the wire, a magnetic field is created and the compass needle is deflected as illustrated below. The net effect of the magnetic field of the Earth and the field produced by the wire is given by B_{net}. The magnetic field of the Earth is B_E and B_W is the field produced by the wire. In this experiment we are trying to measure the magnetic field of the wire as a function of the current through the wire. The needle is deflected by the angle θ.

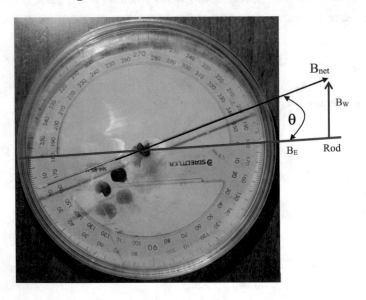

Figure 4.14: Measuring the deflection caused by an applied magnetic field.

The orientation of the rod and compass needle forms a right triangle. Simple trigonometry tells us that

$$\text{Tan}(\theta) = \frac{B_W}{B_E} \, .$$

We can rearrange this equation to solve for B_W. This gives

$$B_E \, \text{Tan}(\theta) = B_W.$$

Now we can measure the strength of the wire's field (B_W) relative the Earth's field (B_E). If we call the Earth's field one unit then the equation becomes

$$\text{Tan}(\theta) = B_W.$$

This means that if $Tan(\theta) = 2.5$, then the field of the wire is 2.5x times the Earth's field or

$$B_W = 2.5B_E.$$

Measuring the angle that the needle is deflected gives us a way to measure the strength of the magnetic field produced by the wire in units of the Earth's magnetic field. Now we can use our compass to explore the equation

$$B_W = \frac{\mu_0 I}{2\pi R} = \text{Tan}(\theta) \, .$$

This equation tells us that the magnetic field produced by the wire should be proportional to the current I through the wire. We can experimentally test the proportionality predicated by this equation. If we graph the tangent of the deflection angle θ as a function of the current through the wire, with the distance held constant, we would expect a straight line, as shown in Figure 4.15.

4.7   DATA ANALYSIS

In this section we will explore the terms in the magnetic field equation by constructing graphs and models of our data. The measure of how well the data fits a model is given by the R^2 value of the graph. If the data points could be perfectly modeled by a linear relationship, the R^2 value would be exactly 1.00. Some of the data points are above the line and some are below the line. When the software constructs a linear regression model, the idea is to find a line that has some data above and some data below the line. The objective is to minimize the sum of the squared differences between the model and the data points.

For the long straight wire, we would expect the strength of the magnetic field to be inversely proportional to the distance with the current held constant. To acquire this data we again measure the deflection of the needle while increasing the distance between the wire and the compass. This is accomplished by using an insulating support to suspend the rod above the compass. In our laboratory we have supports made from PVC pipe for this purpose. The graph shown in Figure 4.16 illustrates the relationship we expect. It is obvious that a straight line linear fit would not model the data.

We can analyze the data for this experiment several different ways. Since we expect a 1/r dependence we can plot the magnetic field strength as a function of the inverse of the distance. Such a plot should produce a good linear fit as shown in Figure 4.17.

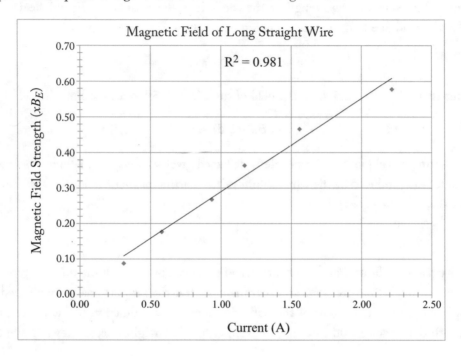

Figure 4.15: Magnetic field strength as a function of current with constant distance.

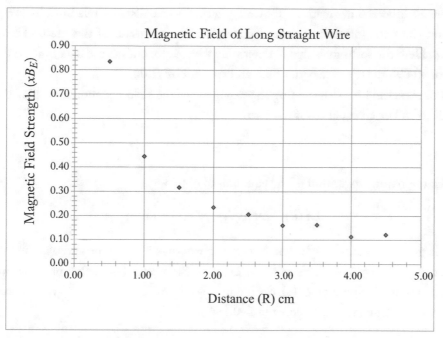

Figure 4.16: Magnetic field strength as a function of distance with constant current.

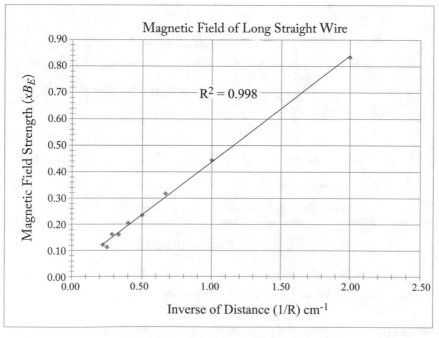

Figure 4.17: Magnetic field strength as a function of the inverse of the distance with constant current.

There are several other methods for verifying the inverse distance relationship. We could plot the data and allow the software to create a power law fit. The results of this method are shown in Figure 4.18. Here the software is free to determine what exponent fits the data best. In this case an exponent of -0.896 fits the data better than an exponent of -1.00.

Another method for determining the exponent of the field equation is to plot the data on logarithmic axes. For an equation of the form

$$B = cr^P,$$

we can take the natural logarithm of the equation to obtain

$$Ln\,(B) = Ln\,(cr^P) = pLn(r) + Ln(c).$$

This has the general form a straight line $Y = mx + b$ where m is the slope and b is the intercept.

We can plot $Ln\,(B)$ on the Y-axis and $Ln(r)$ on the x-axis. For such a plot the slope would be p and the intercept would be $Ln(c)$. Such a plot is shown in Figure 4.19. We can see from the equation that the slope gives an exponent of -0.896.

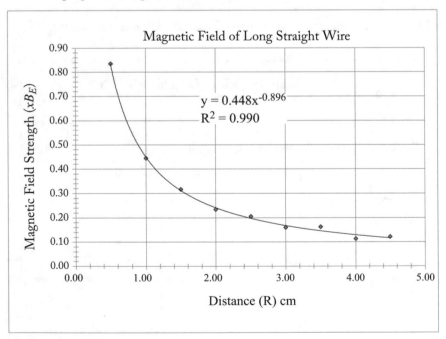

Figure 4.18: Determining the exponent using a power law fit.

You should try to determine the exponent using any of the methods outlined and as directed by your instructor.

In your conclusion section of your lab report discuss the proportionalities we have explored and how they are related to the equation

$$B = \frac{\mu_0 I}{2\pi R} \, .$$

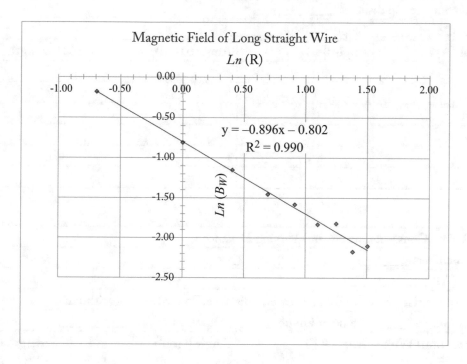

Figure 4.19: Determining the exponent using a logarithmic plot.

4.8 DATA TABLES

Data Table: Constant Distance and Varying Current			
Group members_____			
Wire to compass top distance_____			
R^2 from Graph_____ Slope of Graph_____			
Current (A)	**Angle θ (°)**	**Angle θ (rad)**	**Field Strength (B_E)**

Data Table: Constant Current and Varying Distance			
Group members_____			
Current (A) _____			
Graph type (circle choice): Distance, Inverse distance, Power law, Logarithmic			
R^2 from Graph_____ Slope of Graph_____ Exponent of field_____			
Distance (mm) Wire to Top	**Angle θ (°)**	**Angle θ (rad)**	**Field Strength (BE)**

CHAPTER 5

Copper Mining

铜矿开采铜矿 Tóngkuàng Kāicǎi

5.1 OBJECTIVE

The experiments described in this section are designed to familiarize students with two methods of obtaining copper. In the first experiment we conduct a simple smelting activity using malachite jewelry as the copper ore. An alternative method of known as "wet copper" mining is presented which avoids the use of open flames or high temperature furnaces. These experiments were specifically designed for non-science students. Instructors may feel free to modify or add more complicated chemical principles as they see fit. Copper is, of course, the main ingredient of bronze and was vital to early Chinese metallurgy. In the modern age, copper is used for plumbing, electrical wiring, building materials, coins, jewelry, and even as a dietary mineral.

Chinese vocabulary		
铜 tóng Copper (Cu)	锡 xī Tin (Sn)	铁 tiě Iron (Fe)
青铜 qīngtóng Bronze	铅 qiān Lead (Pb)	硫 liú Sulfur (S)
矿 kuàng Mine; mineral, ore	碳 tàn Carbon (C)	还原 huányuán Reduction
铜矿 tóngkuàng Copper mine	氧 yǎng Oxygen (O)	孔雀石 kǒngquèshí Malachite
开采 kāicǎi to extract or mine	氢 qīng Hydrogen (H)	
氧化铜 yǎnghuàtóng Copper oxide (CuO)		
Materials for Reduction of Malachite		
Malachite stones	Furnace capable of 900°C or Bunsen burner	
Weighing scale	Crucible holder if using Bunsen burner	Crucible lid
Multimeter	Brass bristle wire brush	Charcoal
Fine Tweezer (No. 5)	Crucible, glazed porcelain or Alumina (for melting)	
Crucible tongs	Safety glasses	High temperature gloves

Materials for Wet Copper Mining		
Mass Scale with 10 mg sensitivity		
Graduated cylinder	Plastic tongs or tweezers	Hotplate
Gloves	Filter flask and funnel	Hacksaw
Abrasive cloth	Filter paper	Weighing boat (metal)
Permanent marker	Plastic or wooden scraper or knife	
Steel stock (black weldable)	Deionized water and squeeze bottle	
Plastic container with good seal	Engraver (optional)	

5.2　INTRODUCTION

We present two different methods for obtaining copper, 铜 tóng. One method uses thermal processing of malachite $Cu_2CO_3(OH)_2$, a common copper ore. This pyrometallurgical method does employ a high temperature furnace or open flame and may not be suitable for all learning environments. We also introduce the "wet copper" or hydrometallurgical method which does not require high temperatures and may be suitable for a wide range of student abilities. Performing both experiments will give students the opportunity to compare the two methods and examine their relative strengths and weaknesses. In particular the fuel, labor, and capital equipment costs.

5.3　SMELTING MALACHITE

As we pointed out in Chapter 3, copper can be found in its native form but this is a rare occurrence. There are about a dozen minerals containing a wide range of copper content. Copper minerals span a copper content of about 32% to nearly 89%. Cuprite or copper (I) oxide Cu_2O contains about 88.8% copper and tetrahedrite, a copper antimony sulfosalt $(Cu,Fe)_{12}Sb_4S_{13}$, can contain about 32% copper. Neither of these minerals are found in large quantities. Malachite $Cu_2CO_3(OH)_2$ was known to the ancient Chinese. The Chinese name for malachite is 孔雀石 kǒngquèshí or peacock stone. 孔雀 kǒngquè is peacock and 石 shí means stone. When most people think of a peacock they think of the metallic blue peacocks found in India. There is a green species of peafowl found in South China and other parts of Southeast Asia. This helps us to make sense of the name孔雀石 kǒng què shí. In this experiment we will use malachite chips from an inexpensive necklace as shown in Figure 5.1. You can find such stones in many hobby and craft stores.

　　Copper is, of course, the main component of bronze. In Chapter 3, We discussed the importance of bronze (青铜 qīngtóng) and pointed out that the two main components of bronze are copper (铜 tóng) and tin (锡 xī). For much of China's early history, bronze was the alloy of choice.

This was especially true for use in coins and other items were corrosion resistance was crucial. Iron was used for coins but very few remain today as they would simply rust away. Most bronze used for coins also contained a small percentage of lead (铅 qiān). Tin can be reduced from the tin ore cassiterite much the same way we reduce copper oxide in the malachite experiment. Lead has a low melting point and is easy to work with, but there are health risks that make it less suitable for school use.

Figure 5.1: An inexpensive malachite (孔雀石 kǒngquèshí) necklace used as a source of copper ore.

Copper-bearing minerals have a wide range of copper content and we should calculate what fraction of the mass is expected to be copper. First, we will assume that the malachite is pure so the chemical formula is

$$Cu_2CO_3(OH)_2.$$

We can use the standard atomic weights of the constituent elements to calculate the ratio of the copper mass to the total mass of the mineral. Once we have that ratio, we can apply it to any given starting mass and calculate the theoretical copper yield.

Table 5.1: Formula mass of malachite $Cu_2CO_3(OH)_2$

Element	Quantity Mol of Element	Atomic mass (amu)	Subtotal (amu)
Cu	2	63.546	127.09
C	1	12.011	12.011
O	5	15.999	79.995
H	2	1.0079	2.0158
Formula mass g/mol			221.11

We can calculate the copper yield ratio form the ratio of the copper subtotal to the formula mass:

$$\frac{2\,mol\,Cu}{1\,mol\,Cu_2CO_3(OH)_2} = \frac{127.092\,amu}{221.114\,amu} = 0.575\,or\,57.5\,\%.$$

So for a given mass of malachite a little more than half of the mass should be extracted as copper.

5.3.1   WARNING!

Some people can be allergic or sensitive to copper. Be sure to wear gloves during these experiments. If you have a known copper sensitivity inform your instructor. These experiments require the use of high temperatures or open flames. Perform them only under proper conditions and guidance from and experienced instructor.

5.4   EXPERIMENTAL METHOD—REDUCTION OF MALACHITE

This experimental description uses a furnace as the heat source. You can also use a Bunsen burner if a furnace is not available. Malachite is a copper carbonate and reducing it to copper is relatively simple. We have discussed this in Chapter 3 and we will reproduce that discussion here. If we heat malachite to 350°C or so, it will produce copper oxide (CuO), carbon dioxide (CO_2), and water (H_2O). The reaction is:

$$Cu_2CO_3(OH)_2(s) \xrightarrow{\triangle} 2CuO(s) + CO_2(g) + H_2O(g).$$

As we heat the malachite chips they will expand and crack. This is clearly shown in Figure 5.2. Gasses evolved will also contribute to the fracture. Slow heating in a furnace seems to minimize the destruction of the chips which is useful if you wish to restring the resulting copper to make jewelry. Cover the crucible when heating the chips to decrease the likelihood of being hit by hot material or damaging the inside of the furnace, as shown in Figure 5.3.

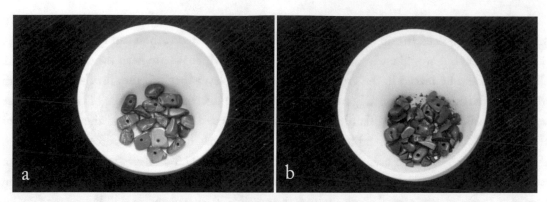

Figure 5.2: Malachite chips before (a) and after (b) heating at 500°C for 20 minutes.

Figure 5.3: Heating malachite chips in a furnace with crucible cover slightly ajar to allow gasses to escape.

Our malachite chips should now just be copper (II) oxide CuO (氧化铜 yǎnghuàtóng). We can remove the chips and weigh them to determine any additional thermal decomposition is necessary. Assuming that we have only copper (II) oxide it is time to reduce this to elemental copper. This next step requires a very high temperature and heating in the presence of carbon. The carbon and oxygen produce carbon dioxide leaving behind elemental copper. The rock chips may contain some other minerals, so there may be impurities as well. The reduction is given by

$$2CuO(s) + C(s) \xrightarrow{\triangle} 2Cu(s) + CO_2(g).$$

This time we will heat the crucible to 900°C and maintain the temperature for 60–90 minutes. First, we need to add some powdered charcoal. Activated carbon can be used but simple charcoal such as one would use in a barbecue grill can be used. Be sure to keep the crucible covered. Charcoal should be placed in the crucible first to make a layer that will prevent the copper from sticking to the bottom. Completely cover the copper oxide in charcoal. The type of crucible we use is very important. Glazed porcelain crucibles are usually rated to 1050°C and the melting point of copper is 1085°C. If you intend to melt the resulting copper, do not use a porcelain crucible, an alumina crucible should be used. Alumina crucibles made of 99.8% Al_2O_3 are suitable for high temperatures can be used up to 1750°C. If you are not using a furnace but heating with a Bunsen burner or torch be very careful about melting the crucible. A cherry-red to dull light-orange color means the crucible is in the 800–900°C range. This is shown in Figure 5.4. Avoid heating to bright yellow or white hot as you will likely melt a glazed porcelain crucible.

Figure 5.4: Heating malachite chips at 900°C in a furnace.

After the malachite chips have been heated, remove them from the furnace and inspect their color.

Figure 5.5: Chips after heating with charcoal.

Some of the chips in Figure 5.5 have a reddish color. This indicates that they have not been fully reduced. A different oxide has been formed. Copper (I) oxide Cu_2O has a distinct red color. These chips can be reheated in the furnace with additional charcoal. Your chips will not necessarily look like a nice shiny metal and may require a little cleaning. They can be cleaned in a rock tumbler or with a brass bristle brush. You may wish to save the malachite chips and restring them to make your own jewelry.

Figure 5.6: Malachite chips after reduction and cleaning.

5.5   EXPERIMENTAL PROCEDURE—REDUCTION OF MALACHITE

These instructions assume that you will use a Bunsen burner as the heat source. If using a furnace still record the same data

  1. Select several grams of malachite chips. Try to choose those of uniform color.

Figure 5.7: Malachite chips.

Record the initial mass of the chips in your data table.

2. Weigh an empty crucible and place the malachite chips in the crucible. Weigh the crucible again with chips and record this in your data sheet.

3. It is important to note that a glazed porcelain crucible can be damaged if you heat it above 1000°C. Roast the chips at about 500°C. This is accomplished by placing the crucible in the cooler part of the Bunsen burner flame which is shown in Figure 5.8. Make sure the crucible is covered! As the chips heat up they may expand and crack. You can be burned by hot chip fragments so keep the lid on for protection.

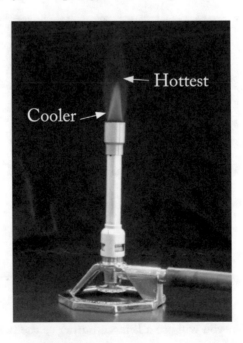

Figure 5.8: Bunsen burner flame and different temperature regions.

A Bunsen burner flame has different temperature regions. The flame generally has an inner cone and an outer cone. The hottest part of the flame is just beyond the tip of the inner cone. To achieve a lower temperature place the crucible in the cooler part of the flame below the tip of the inner cone and closer to the barrel. Heat the chips for about 15–20 minutes.

Figure 5.9: Heating malachite chips in cooler part of flame.

4. Use tongs to move the crucible out of the flame and place it on a fire brick or other non-flammable material to cool down.

5. Weigh the crucible with the chips in it and record this in your data sheet.

6. Determine the mass of the remaining product (2CuO) and the mass lost due to roasting. If the crucible shows signs of damage, remove the chips and weigh them to determine the mass of the remaining product.

7. From the formula for malachite and the formula for the solid product (2CuO) calculate the theoretical mass ratio (mass product/mass malachite).

8. From the theoretical mass ratio and the starting mass of malachite calculate the theoretical mass of the product. Compare these two masses and show the percent difference with respect to the theoretical value.

9. Remove one of the larger chips and use a multimeter to check if the chip is conducting. This is shown in Figure 5.10. Measure the electrical resistance and note this in your laboratory report.

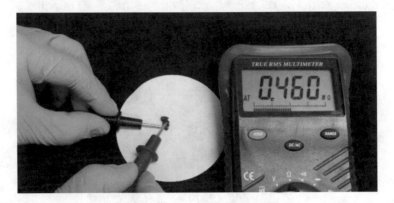

Figure 5.10: Measuring the electrical resistance of a chip that has been roasted but not yet reduced to elemental copper.

Now it is time to reduce the copper (II) oxide to elemental copper.

Remember the crucible warning. This time you will heat the crucible until the bottom is cherry red or about 900°C if you are using a furnace. This stage will require 60–90 minutes.

10. Pour some charcoal into the crucible and make a layer to keep the copper oxide product from sticking to the bottom of the crucible. Add the roasted product and cover it with more charcoal. Fill the crucible nearly to the top with charcoal.

11. Weigh the crucible with all the contents and determine how much charcoal you added. Record this in your data sheet. Be sure to cover the crucible.

12. Heat crucible to a light orange color or 900°C for 60 minutes.

13. Once the crucible has cooled to near room temperature, use tongs to remove the lid. Examine the contents and look for any dull brick red to bright red traces of copper (I) oxide.

14. If you do see evidence of red copper (I) oxide, heat your product under charcoal again to reduce it. You may wish to separate out the only the red chips for this reheating step.

Figure 5.11: Heating in the hotter part of the flame for reduction.

15. Once you are satisfied that you have reduced all of the malachite chips, spread the contents out onto a clean sheet of white paper. Separate out all of the reduced product which should be elemental copper. Be sure to get the very small broken fragments.

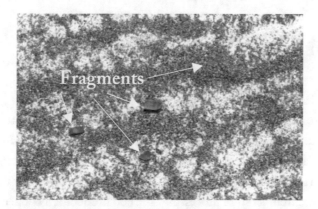

Figure 5.12: Harvest all small fragments from among the charcoal ash.

16. Weigh all of the reduced material and record this in your data sheet.

17. Using the copper yield ratio we previously determined, calculate the theoretical amount of copper that should be produced from the original malachite. Calculate the percent difference of the copper yield compared to the theoretical value.

18. Clean the copper chips. You can use the brass bristle brush on the larger chips. You should be able to see the copper.

19. Copper is a good conductor. Use the multimeter again to check the conductivity of one of the larger chips. Report this in your lab report.

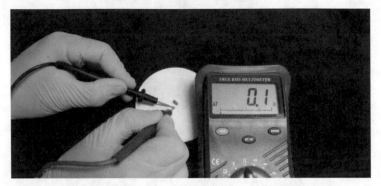

Figure 5.13: Measuring the electrical resistance of a chip that has been reduced to elemental copper.

Many of the copper chips will still have a hole in them. If you want you can string them together to make some sort of jewelry. Warning! Copper is NOT hypoallergenic and you could have a reaction to copper jewelry. It may turn you skin green or cause some other allergic reaction. It is wise to apply a thin transparent layer of nail polish over the copper to prevent skin contact.

5.6   WET COPPER METHOD

5.6.1   INTRODUCTION

We have already discussed some aspects of the wet copper method in Chapter 3. This technique of mining copper is sometimes called copper precipitation or cementation. Many different Chinese terms are used to describe the process. A common term is 胆水浸法 dǎnshuǐjìnfǎ, which is sometimes translated as vitriol water steeping method. It is a hydrometallurgical method and as such it does not require any high temperature processing. The end product is a copper mud. If an appropriate heat source is available, the mud can be melted to form a copper ingot, but that is optional. At wet copper mines, the mud is stored in sacks that are later shipped off to a smelter. The copper is not strongly adhered to the steel surface and it is easily removed. To harvest the copper one simply needs to scrape or wash off the surface. These steps are shown in Figure 5.14.

Figure 5.14: Rinsing copper mud form scrap metal. Photographs courtesy of Alexander Jost.

At the smelter the mud undergoes additional purification and a final melting in to an ingot. In our experiment we are mainly concerned in producing the copper mud which can be dried to form powder.

In an actual wet copper production facility there is often an inflow of fresh solution which meanders over iron or iron containing metal such a steel. A simple trench or flume with scrap steel can become an effective "mine" as Figure 5.15 demonstrates. In this configuration water just continuously flows across the scrap metal.

Figure 5.15: Wet copper "mines" can simply be flumes filled with scrap metal. Photographs courtesy of Alexander Jost.

In this experiment we will reproduce the wet copper method but will not use a continuous source of fresh solution. We will start out with a known volume of solution and examine how the rate of copper deposition changes with time. As we described in Chapter 3, the wet copper process is an example of a displacement reaction in which the iron (Fe) and copper (Cu) atoms exchange places. Iron from the steel bars goes into solution to form iron sulfate and copper atoms from the copper sulfate solution deposits or plates out onto the steel bars. The chemical equation describing the process is

$$Fe(s) + CuSO_4(aq) \xrightarrow{\triangle} Cu(s) + FeSO_4(aq).$$

The solution we will use is Copper Sulfate Pentahydrate $CuSO_4 \cdot 5(H_2O)$, so we need to calculate the formula mass.

Table 5.2: Formula mass for Copper Sulfate Pentahydrate $CuSO_4$ 5(H_2O)			
Element	Quantity Mol of Element	Atomic mass (amu)	Subtotal (amu)
Cu	1	63.546	63.546
S	1	32.059	32.059
O	9	15.999	143.99
H	10	1.0079	10.079
Formula mass g/mol			249.68

The mass fraction of this that is copper is given by the ratio of the formula masses:

$$Fraction\ Cu = \frac{63.546\ g/mol}{249.68\ g/mol} = 0.25451\ or\ about\ 25.5\%.$$

If we know the starting concentration and the volume of the solution, we can use this factor to calculate the theoretical copper yield. The concentration of the solution will be determined by the laboratory instructor. We do not want the experiment to take hours to complete so the concentration should be high enough to collect easily measurable amounts of copper in time periods of a few minutes. A 10% w/v solution works well for this purpose. Just as a reminder, the notation w/v means weight by volume and is the mass in grams of the dissolved solute divided by the volume in milliliters of the solution.

5.6.2   EXPERIMENTAL METHOD—WET COPPER

We will then "minc" this known starting solution by allowing it to react with steel bars in a closed container. Be careful not to spill or lose any solution during the mining process. It is helpful to have a few paper towels ready. You can determine the mass of these towels beforehand and then if you have to wipe up any spills measure their mass again to determine how much solution was lost.

Before placing the steel bars in solution, you should clean, weigh, and mark each bar. Your bars may have rust or some other coating such as oil. Use abrasive cloth or sandpaper to make sure you start with a nice clean surface. It is helpful to mark the steel bars so you can identify them and record their initial mass. Use an electric engraver or hard steel point to scratch an identifying number on the bar. Your instructor may have you keep track of the mass loss and amount of iron that goes into solution. Steel is not pure iron, so this number might not correctly reflect the mass lost by the steel bars. Pure iron is expensive and steel makes a good substitute.

Once the steel bars are placed in solution close the lid and gently agitate the tank. Try to agitate the tank with the same vigor throughout the entire experiment. After a predetermined amount of time remove the bars from the tank and let them drip off into the box. You can use plastic tweezers or some other unreactive material to remove the bars. When you remove the bars form the tank be careful not to damage the copper coating. Some may flake off and fall into the tank. At the end of the experiment, be sure to recover any material remaining in the tank. This will contribute to the overall yield.

You will now capture the copper that deposited onto the steel bar. We will use filter paper to accomplish this. Before you begin this part of the experiment, be sure to determine the mass of several circles of filter paper. Number each filter and record the mass in your data table. Weigh the paper after you mark it. You may need to prepare ten or more circles of filter paper.

Before we begin harvesting any copper, we should realize that filter paper is generally hygroscopic and can absorb humidity from the air. If you wish to conduct this experiment to high precision it will be necessary to burn the filter paper and then measure the mass of the copper captured by the filter. This is a good analytical technique to learn. However, this experiment can be conducted by simply weighing the copper on the filter paper after it has been dried. This will introduce some small uncertainty but you should be able to achieve results within a few percent of the theoretical yield.

In order to determine the amount of copper deposited onto the steel bars, we scrape the coating with a wooden or plastic instrument creating small copper flakes. These flakes are captured on filter paper by rinsing them off with distilled water. Before use, the filter paper is labeled and weighed. When we weigh the dry filter paper it has actually absorbed some humidity from the air and this contributes to its weight. After rinsing the copper flakes off of the bar, the wet filter paper must be dried out.

5.6.3   CHECKING THE FILTER PAPER WEIGHT

If you weigh the filter paper while it is still hot, the apparent weight change is misleading. The hot filter paper will have become dehydrated and weigh less than it did initially. Once the paper is dry, allow it to equilibrate to room temperature and humidity. After it has reached equilibrium with the room, then weigh the dry filter paper which is now coated with copper flakes. Different grades of filter paper have different characteristics. **Before** conducting the experiment test your filter paper. Weigh a circle of filter paper at room temperature. Then heat it on a hot plate for 10 or 15 minutes to dehydrate it. Remove the paper from the hotplate and immediately weigh it while it is still hot. Is there a change in weight? Now allow the filter paper to come to equilibrium with its surroundings. Weigh the paper again and see if it has returned to the initial weight. If the weight is different, weigh it again and see if it recovers back to the initial weight in time. Be sure that you understand how long the filter paper needs to equilibrate with its surroundings and what degree of uncertainty is expected. Comment on this point in your laboratory report. We will calculate the experimental yield and compare that to the theoretical yield. Uncertainties introduced by the weighing procedure must be accounted for in analyzing the results.

Hold the steel bars over a funnel with filter paper and gently scrape the entire surface with a wooden or plastic instrument. You do not want to scrape off any of the steel, only the copper coating. As you scrape, wash the dislodged material onto the filter with some distilled or deionized water. Allow all of the rinse water to pass through the filter before you remove the filter paper. Wet filter paper can easily be torn. Be careful not to rip or puncture the paper with the steel bar; otherwise your valuable copper will end up in the filter flask and you will have to recover it.

In this step we will measure the mass of the collected copper. Use a temperature-resistant weighing boat such as aluminum. Determine the mass of the boat before you place the filter paper in it. The boat helps to capture any copper that may spill or slip off of the filter paper. Place the wet filter paper in a temperatur-resistant weighing boat and heat on a hot plate. The hot plate will boil off any remaining water and dry out the filter paper. You can fit several wet filters on the surface of the hot plate and allow them to dry out simultaneously.

Now repeat the process several more times. You should be able to notice that the copper yield is not always the same. Eventually you will have to increase the time that the bars are exposed to solution to get a measurable amount of copper. This is an important point that we want to explore. Naively, we might think that if we double the time that the bars are in solution we should get double the amount of copper. Is this true? Of course, we are limited by the amount of copper that is initially in the solution so we cannot expect more than that amount. Even below that limit, the rate of the reaction will not be constant. Discuss this point in your laboratory report.

5.6.4 EXPERIMENTAL PROCEDURE—WET COPPER METHOD

1. Obtain copper sulfate solution of known concentration and pour known volume into the plastic tank, as shown in Figure 5.16. Record the volume and concentration in your data table. Place the lid on the tank to prevent any spills. Also start the hotplate so it will be warmed up when you need it. It should be warm enough to evaporate water from the wet filter paper.

Figure 5.16: Fill the tank with known amount of solution.

2. Now we prepare the steel bars as shown in Figure 5.17. You will need to mark the bars so you can identify which is which and record their mass. Use a scribe or engraver to make the bar. Alternatively, you can use a saw to cut a notch in the side. The surface does not have to be smooth; use a coarse grit.

Figure 5.17: Clean the numbered steel bar.

3. Weigh each bar and record their masses in your data table. Also, weigh a few dry paper towels to use for cleanup. The mass of the wet towel can be used to determine the volume of liquid lost. You should also label and weigh several circles of filter paper. Record the dry mass of each filter paper in your data table. For each filter paper also weigh a temperature resistant boat. Record all masses in you data table.

Figure 5.18: Weigh each bar and several circles of filter paper.

4. Place the bars into the tank, as shown in Figure 5.19. Plastic tweezers or tongs will be helpful for this step. Do not splash out any liquid. If you do splash out solution, wipe it up with a paper towel of known mass and then estimate the volume lost or refill the tank with a known volume. Now gently agitate the tank for several minutes. A

rocking motion is sufficient. Be sure to record the time that the bars are in the solution and also keep a running total.

Figure 5.19: Place bars in tank and gently agitate.

5. Use tongs or tweezers to remove the bars and let them drip into the tank. Do not let one bar sit very long in the tank while you are removing the other. We are trying to keep the exposure time the same for each. You can place them diagonally across the tank, as illustrated in Figure 5.20. There should be a nice coating of copper on the surface.

Figure 5.20: Remove bars and let them drip into the tank.

6. Now it is time to harvest the copper we deposited. Use one of the filter papers of known dry mass to capture the copper. Gently abrade all surfaces of the bar with a non-metallic instrument such as a wooden scraper of plastic knife. You just want to loosen the copper and not scrape off the steel. Then rinse the dislodged copper off the surface and onto the filter paper, as shown in Figure 5.21. Repeatedly scrape and

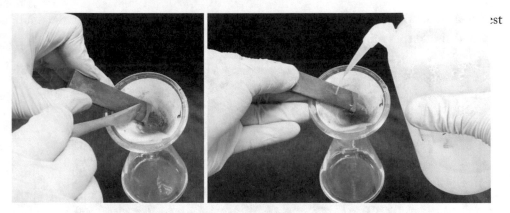

Figure 5.21: Scrape and rinse the copper onto a numbered circle of filter paper with known dry mass.

7. Allow the rinse water to drain through the filter and then remove it from the funnel. Placc the wet filter into a temperature-resistant weighing boat of known mass. If any copper from the filter paper leaks out into the boat, you can determine its mass from the known mass of the dry boat. Place the wet filter paper and boat onto a hotplate and let the filter dry out, as shown in Figure 5.22.

Figure 5.22: Dry the wet filter paper on a hotplate.

As you progress through the experiment you will have a crowd of filters drying out on the hotplate. Once they seem dry, remove them from the hot plate and let them equilibrate with the room before you weigh them.

8. Place the steel bars back in the tank for a pre-determined amount of time. Harvest copper from the bars repeatedly. If you notice that very little copper is deposited try doubling or tripling the time. As you remove copper from the solution it will take longer to deposit a noticeable amount. For each harvest, be sure to have a pre-weighed boat and filter paper.

9. When you decide that you have reached the limit where very little is deposited in about 30 minutes, you have probably harvested most of the available copper. As you have been handling the steel bars, you may have caused some copper flakes to fall off into the bottom of the tank. Use one more filter paper to capture the remains from the bottom of the tank.

In the next section we will consider some points of analysis that you should include in your laboratory report.

5.7   ANALYSIS

Now it is time to see how good of a miner you are. Just what does that mean? You might think that being a good miner means that you got every last milligram of copper out of the solution. Would that be cost effective? Suppose you had to pay an employee an hourly rate. Would it be cost effective to spend an extra half hour just to get that last little bit or would it be more cost effective to work with fresh solution? In Chapter 3, we quoted the price of copper at the time this was written. You can easily look up the market price for copper. With the small scale of experiment you may have harvested a few U.S. cents worth of metal. You should reflect on this in your laboratory report and consider how such an operation may be run cost effectively. Describe how you would set up such a facility and what factors must be considered to make it cost effective.

Here is another question to consider in your lab report. Suppose we harvest 0.30 g of copper in 5 minutes with a continuous flow of fresh solution flowing over the steel. This might be a setup similar to what is shown in Figure 5.15. Will we be able to harvest ten times as much copper if we just let the steel bar soak for ten times as long? What happens to the rate of the reaction when the bar is thickly coated with copper?

In our experiment we did not have a continuous flow of fresh solution. We started with a known concentration and cleaned off the steel bars periodically. From your data plot a graph of the amount of copper deposited as a function of the cumulative time that the bars were in solution. You should obtain a curve that looks similar to Figure 5.23.

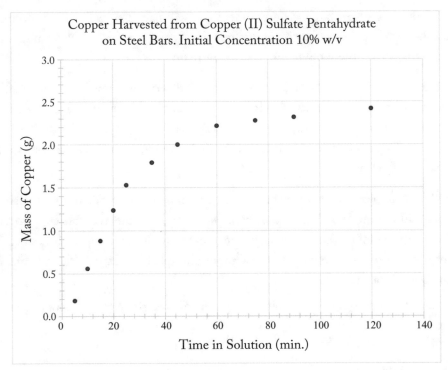

Figure 5.23: Copper harvested as a function of exposure time.

Is the curve linear? What does the shape of the curve imply about attempting to harvest say 95% of the theoretically available copper? Based on the concentration and volume of the starting solution, calculate the amount of the theoretically available copper. Compare the total amount of copper you harvested to the theoretical amount of available copper. Calculate the percent difference between the experimental and theoretical values.

Author Biography

Matt Marone is an Associate Professor of Physics at Mercer University in Macon Georgia where he teaches Physics, Astronomy, and Asian Studies. He received his Ph.D. and M.S. degrees from Clemson University in the area of experimental solid state physics and a B.S. degree in physics from the Rochester Institute of Technology. In the early 1990s, he worked as a post-doctoral researcher at the Chinese Academy of Sciences in Beijing. At Mercer he teaches a wide range of physics classes including several specialized classes in observational astronomy, acoustical foundations of music and ancient Chinese science. In addition to his academic research, he is also active in the area of space resources. This work involves the extraction of oxygen and metals from the regolith of the Moon, Mars, and asteroids. As a member of the Society for Georgia Archeology, he assists in the study and interpretation of Georgia's historic and prehistoric archaeological heritage.